R&AW and Civil Intelligence
A Factful Analysis

R&AW and Civil Intelligence
A Factful Analysis

Dr. C.P. Thakur

NEHA PUBLISHERS & DISTRIBUTORS
DELHI

Publisher
NEHA PUBLISHERS & DISTRIBUTORS
4832/24,Prahlad Lane,S-207 Ansari
Road, Daryaganj, Delhi-110002
Ph.: 43570976, 23278261
Email: nehapubdistributors@gmail.com

Edition: 2022

ISBN: 978-93-80318-84-4

Laser Typesetting
JEE-VEE Graphics, Delhi

Price: 1295/-

Printed
Vikas Computers, Delhi

Preface

RAW only oversees foreign intelligence where as internal part is taken care by Indian IB. Only in few exception cases RAW was given responsibility of a domestic issue.

Research and Analysis Wing (RAW or R&AW) is the primary external intelligence agency of the Republic of India. It was formed in September 1968 under the helmsman-ship of its first Director, Rameshwar Nath Kao. Its creation was necessitated post the Sino-Indian War 1962 and Indo-Pakistani War of 1965 which posed various gaps in intelligence gathering undertaken by Intelligence Bureau (which then handled both internal and external intelligence). This convinced the Government of India that a specialised, independent agency was required for competent external intelligence gathering.

The primary function of R&AW is collection of external intelligence and counter-terrorism. In addition, it is responsible for obtaining and analysing information about foreign governments, corporations and persons to advise Indian policymakers. R&AW is an effective and one of the primary instrument of India's national power. It is also involved in the security of India's nuclear programme.

R&AW personnel are called "Research Officers" instead of the traditional "agents". There is a sizeable number of female officers in R&AW even at the operational level. In recent years, R&AW has shifted its primary focus from Pakistan to China and have started operating a separate desk for this purpose.

Though each agency is charged with its own mission, the government has provided a means to facilitate the sharing of information between members of the intelligence community.

—Editor

Contents

1

RAW (Research & Analysis Wing): India's CIA

The Research and Analysis Wing famously known as RAW, is India's foreign intelligence agency. RAW has become an effective instrument of Indian national power, and has assumed a significant role in carrying out India's domestic and foreign policies. RAW has engaged in spying against Pakistan and other neighbouring countries. It has enjoyed the backing of successive Indian governments in these efforts. Working directly under the Prime Minister, the structure and operations of the Research & Analysis Wing are kept secret from Parliament.

Founded in 1968, RAW focused largely on Pakistan. Its formation was initially motivated by reports of Pakistan supplying weapons to Sikh militants, and providing shelter and training to guerrillas in Pakistan.

Most recently, RAW has gained attention for providing the US with intelligence on Al-Qaeda and Taliban targets for the war on terrorism in Afghanistan. Maps and photographs of terrorist training camps in Afghanistan and Pakistan, along with other evidence implicating Osama bin Laden in terrorist attacks, were given to US intelligence officials.

THE STRUCTURE AND FUNCTION OF RAW

Not much is known regarding the structure of RAW, say experts. The organization started with 250 people and about $400,000. It has since expanded to several thousand personnel, but its staffing and budget remain secret. However, an estimate by the U.S.-based Federation of American Scientists suggests that in 2000, RAW had about eight to ten thousand agents and a budget that experts place at $145 million. Unlike the United States' Central Intelligence Agency (CIA) or Britain's MI6, RAW reports directly to the prime minister instead of the Ministry of Defense. The chief of RAW is designated secretary (research) in the Cabinet Secretariat, which is part of the prime minister's office. Some officers of RAW are members of a specialized service, the Research and Analysis Service, but several officers also serve on deputation from other services such as the Indian Police Service.

RAW had two priorities after its formation, writes B. Raman, a former RAW official, in the 2007 book, *The Kaoboys of R&AW: Down Memory Lane*. The organization worked to strengthen its capability for intelligence gathering on Pakistan and China and for covert action in East Pakistan (now Bangladesh). Some experts say that RAW's efforts in East Pakistan, which was created from the partition of the Indian state of Bengal and completely separated from the rest of Pakistan, was aimed at fomenting independence sentiment. Over time, RAW's objectives have broadened to include:

- Monitoring the political and military developments in adjoining countries, which have direct bearing on India's national security and in the formulation of its foreign policy.
- Seeking the control and limitation of the supply of military hardware to Pakistan, mostly from European countries, the United States, and China.

Experts disagree on the amount of influence RAW asserts on India's foreign policy. Sumit Ganguly, a professor of political science at Indiana University, says the agency has no influence on foreign policy. However, Dipankar Banerjee, a retired army official and founding director of the Institute of Peace and Conflict Studies, a New Delhi-based think tank, says the head of RAW has direct access to the head of state, to whom he provides input and analysis.

From the early days, RAW had a secret liaison relationship with the Mossad, Israel's external intelligence agency. The main purpose was to benefit from Israel's knowledge of West Asia and

North Africa, and to learn from its counterterrorism techniques, say experts.

RAW: AN INSTRUMENT OF INDIAN IMPERIALISM

The Research and Analysis Wing (RAW), created in 1968, has assumed a significant status in the formulation of India's domestic and foreign policies, particularly the later. Working directly under the Prime Minister, it has over the years become and effective instrument of India's national power. In consonance with Kautilya's precepts, RAW's espionage doctrine is based on the principle of waging a continuous series of battles of intrigues and secret wars.

RAW, ever since its creation, has always been a vital, though unobtrusive, actor in Indian policy-making apparatus. But it is the massive international dimensions of RAW operations that merit a closer examination. To the credit of this organization, it has in very short span of time mastered the art of spy warfare. Credit must go to Indira Gandhi who in the late 1970s gave it a changed and much more dynamic role. To suit her much publicized Indira Doctrine, (actually India Doctrine) Mrs. Gandhi specifically asked RAW to create a powerful organ within the organization which could undertake covert operations in neighbouring countries. It is this capability that makes RAW a more fearsome agency than its superior KGB, CIA, MI-6, BND and the Mossad.

Its internal role is confined only in monitoring events having bearing on the external threat. RAW's boss works directly under the Prime Minister. An Additional Secretary to the Government of India, under the Director RAW, is responsible for the Office of Special Operations (OSO), intelligence collected from different countries, internal security (under the Director General of Security), the electronic/technical section and general administration. The Additional Secretary as well as the Director General of Security is also under the Director of RAW. DG Security has two important sections: the Aviation Research Center (ARC) and the Special Services Bureau (SSB). The joint Director has specified desks with different regional divisions/areas (countries):

Area one. Pakistan: Area two, China and South East Asia: Area three, the Middle East and Africa: and Area four, other countries.

Aviation Research Center (ARC) is responsible for interception, monitoring and jamming of target country's communication systems. It has the most sophisticated electronic equipment and also a substantial number of aircraft equipped with state-of- the art eavesdropping devices. ARC was strengthened in mid-1987 by the addition of three new aircraft, the Gulf Stream-3. These aircraft can reportedly fly at an altitude of 52,000 ft and has an operating range of 5000 kms. ARC also controls a number of radar stations located close to India's borders. Its aircraft also carry out oblique reconnaissance, along the border with Bangladesh, China, Nepal and Pakistan.

RAW having been given a virtual carte blanche to conduct destabilization operations in neighbouring countries inimical to India to seriously undertook restructuring of its organization accordingly. RAW was given a list of seven countries (Bangladesh, Sri Lanka, Nepal, Sikkim, Bhutan, Pakistan and Maldives) whom India considered its principal regional protagonists. It very soon systematically and brilliantly crafted covert operations in all these countries to coerce, destabilize and subvert them in consonance with the foreign policy objectives of the Indian Government.

RAW's operations against the regional countries were conducted with great professional skill and expertise. Central to the operations was the establishment of a huge network inside the target countries. It used and targeted political dissent, ethnic divisions, economic backwardness and criminal elements within these states to foment subversion, terrorism and sabotage. Having thus created the conducive environments, RAW stage-managed future events in these countries in such a way that military intervention appears a natural concomitant of the events. In most cases, RAW's hand remained hidden, but more often that not target countries soon began unearthing those "hidden hand". A brief expose of RAW's operations in neighbouring countries would reveal the full expanse of its regional ambitions to suit India Doctrine.

Bangladesh

Indian intelligence agencies were involved in erstwhile East Pakistan, now Bangladesh since early 1960s. Its operatives were in touch with Sheikh Mujib for quite some time. Sheikh Mujib went to Agartala in 1965. The famous Agartala case was unearthed in 1967. In fact, the main purpose of raising RAW in 1968 was to organise covert operations in Bangladesh. As early as in 1968, RAW was given a green signal to begin mobilising all its resources for the impending surgical intervention in erstwhile East Pakistan. When in July 1971 General Manekshaw told Prime Minister Indira Gandhi that the army would not be ready till December to intervene in Bangladesh, she quickly turned to RAW for help. RAW was ready. Its officers used Bengali refugees to set up Mukti Bahini. Using this outfit as a cover, Indian military sneaked deep into Bangladesh. The story of Mukti Bahini and RAW's role in its creation and training is now well-known. RAW never concealed its Bangladesh operations.

Interested readers may have details in Asoka Raina's Inside RAW: the story of India's secret service published by Vikas Publishing House of New Delhi. The creation of Bangladesh was masterminded by RAW in complicity with KGB under the covert clauses of Indo-Soviet Treaty of Friendship and Co-operation (adopted as 25-year Indo-Bangladesh Treaty of Friendship and Co-operation in 1972).

RAW retained a keen interest in Bangladesh even after its independence. Mr. Subramaniam Swamy, Janata Dal MP, a close associate of Morarji Desai said that Rameswar Nath Kao, former Chief of RAW, and Shankaran Nair upset about Sheikh Mujib's assassination chalked a plot to kill General Ziaur Rahman. However, when Morarji Desai came into power in 1977 he was indignant at RAW's role in Bangladesh and ordered operations in Bangladesh to be called off; but by then RAW had already gone too far. General Zia continued to be in power for quite some time but he was assassinated after Indira Gandhi returned to power, though she denied her involvement in his assassination.

RAW was involved in training of Chakma tribals and Shanti Bahini who carry out subversive activities in Bangladesh. It has also unleashed a well-organized plan of psychological warfare, creation of polarisation among the armed forces, propaganda by false allegations of use of Bangladesh territory by ISI, creation of dissension's among the political parties and religious sects, control of media, denial of river waters, and propping up a host of disputes in order to keep Bangladesh under a constant political and socio-economic pressure.

Sikkim and Bhutan

Sikkim was the easiest and most docile prey for RAW. Indira Gandhi annexed the Kingdom of Sikkim in mid-1970s, to be an integral part of India. The deposed King Chogyal Tenzig Wangehuck was closely followed by RAW's agents until his death in 1992.

Bhutan, like Nepal and Sikkim, is a land-locked country, totally dependent on India. RAW has developed links with members of the royal family as well as top bureaucrats to implements its policies. It has cultivated its agents amongst Nepalese settlers and is in a position to create difficulties for the Government of Bhutan. In fact, the King of Bhutan has been reduced to the position of merely acquiescing into New Delhi's decisions and go by its dictates in the international arena.

Sri Lanka

Post-independence Sri Lanka, inspire of having a multi-sectoral population was a peaceful country till 1971 and was following independent foreign policy. During 1971 Indo-Pakistan war despite of heavy pressure from India, Sri Lanka allowed Pakistan's civil and military aircraft and ships to stage through its air and sea ports with unhindered re-fuelling facilities. It also had permitted Israel to establish a nominal presence of its intelligence training set up. It permitted the installation of high powered transmitter by Voice of America (VOA) on its territory, which was resented by India.

It was because of these 'irritants' in the Indo-Sri Lanka relations that Mrs Indira Gandhi planned to bring Sri Lanka into the fold of the so-called Indira Doctrine (India Doctrine) Kao was told by Gandhi to repeat their Bangladesh success. RAW went looking for militants it could train to destabilize the regime. Camps were set up in Tamil Nadu and old RAW guerrillas trainers were dug out of retirement. RAW began arming the Tamil Tigers and training them at centres such as Gunda and Gorakhpur. As a sequel to this ploy, Sri Lanka was forced into Indian power-web when Indo-Sri Lanka Accord of 1987 was singed and Indian Peace-Keeping-Force (IPKF) landed in Sri Lanka.

The Ministry of External Affairs was also upset at RAW's role in Sri Lanka as they felt that RAW was still continuing negotiations with the Tamil Tiger leader Parabhakran in contravention to the Indian government's foreign policy. According to R Swaminathan, (former Special Secretary of RAW) it was this outfit which was used as the intermediary between Rajib Gandhi and Tamil leader Parabhakaran. The former Indian High Commissioner in Sri Lanka, J.N. Dixit even accused RAW of having given Rs. five corore to the LTTE. At a later stage, RAW built up the EPRLF and ENDLF to fight against the LTTE which turned the situation in Sri Lanka highly volatile and uncertain later on.

Maldives

Under a well-orchestrated RAW plan, on November 30 1988 a 300 to 400-strong well trained force of mercenaries, armed with automatic weapons, initially said to be of unknown origin, infiltrated in boats and stormed the capital of Maldives. They resorted to indiscriminate shooting and took high-level government officials as hostages. At the Presidential Palace, the small contingent of loyal national guards offered stiff resistance, which enabled President Maumoon Abdul Gayoom to shift to a safe place from where he issued urgent appeals for help from India, Pakistan, Sri Lanka, Britain and the United States.

The Indian Prime Ministe Rajiv Gandhi reacted promptly and about 1600 combat troops belonging to 50 Independent Para-Brigade in conjunction with Indian Naval units landed at Male under the code-name Operation Cactus. A number of IAF transport aircraft, escorted by fighters, were used for landing personnel, heavy equipment and supplies. Within hours of landing, the Indian troops flushed out the attackers form the streets and hideouts. Some of them surrendered to Indian troops, and many were captured by Indian Naval units while trying to escape along with their hostages in a Maldivian ship, Progress Light. Most of the 30 hostages including Ahmed Majtaba, Maldives Minister of Transport, were released. The Indian Government announced the success of the Operation Cactus and complimented the armed forces for a good job done.

The Indian Defense Minister while addressing IAF personnel at Bangalore claimed that the country's prestige has gone high because of the peace-keeping role played by the Indian forces in Maldives. The International Community in general and the South Asian states in particular, however, viewed with suspicious the over-all concept and motives of the operation. The western media described it as a display of newly-acquired military muscle by India and its growing role as a regional police. Although the apparent identification of the two Maldivian nationals could be a sufficient reason, at its face value, to link it with the previous such attempts by the mercenaries, yet other converging factors, indicative of involvement of external hand, could hardly be ignored. Sailing of the mercenaries from Manar and Kankasanturai in Sri Lanka, which were in complete control of IPKF, and the timing and speed of the Indian intervention proved their involvement beyond any doubt.

Nepal

Ever since the partition of the sub-continent India has been openly meddling in Nepal's internal affairs by contriving internal strife and conflicts through RAW to destabilize the successive legitimate governments and prop up puppet regimes which would be more amenable Indian machinations. Armed insurrections were sponsored and abetted by RAW and later requests for military assistance to control these were managed through pro-India leaders. India has been aiding and inciting the Nepalese dissidents to collaborate with the Nepali Congress. For this they were supplied arms whenever the King or the Nepalese Government appeared to be drifting away from the Indian dictates and impinging on Indian hegemonic designs in the region. In fact, under the garb of the so-called democratization measures, the Maoists were actively encouraged to collect arms to resort to open rebellion against the legitimate Nepalese governments. The contrived rebellions provided India an opportunity to intervene militarily in Nepal, ostensibly to control the insurrections which were masterminded by the RAW itself. It was an active replay of the Indian performance in Sri Lanka and Maldives a few years earlier. RAW is particularly aiding the people of the Indian-origin and has been providing them with arms and ammunition. RAW has also infiltrated the ethnic Nepali refugees who have been extradited by Bhutan and have taken refuge in the eastern Nepal. RAW can exploit its links with these refugees in either that are against the Indian interest. Besides the Nepalese economy is totally controlled by the Indian money lenders, financiers and business mafia.

Afghanistan

Since December 1979, throughout Afghan War, KGB, KHAD (WAD) (former Afghan intelligence outfit) and RAW stepped up their efforts to concentrate on influencing and covert exploitation of the tribes on both sides of the Pakistan-Afghanistan border. There was intimate co-ordination between the three intelligence agencies not only in Afghanistan but in destabilization of Pakistan through subversion and sabotage plan related to Afghan refugees and mujahideen, the tribal belt and inside Pakistan. They jointly organized spotting and recruitment of hostile tribesmen and their training in guerrilla warfare, infiltration, subversion, sabotage and establishment of saboteur force/terrorist organizations in the pro-Afghan tribes of Pakistan in order to carry out bomb explosions in Afghan refugee camps in NWFP and Baluchistan to threaten and pressure them to return to Afghanistan. They also carried out bomb blasts in populated areas deep inside Pakistan to create panic and hatred in the minds of locals against Afghan refugee mujahideen for pressurizing Pakistan to change its policies on Afghanistan.

Pakistan

Pakistan's size, strength and potential have always overawed the Indians. It, therefore, always considers her main opponent in her expansionist doctrine. India's animosity towards Pakistan is psychologically and ideologically deep-rooted and unassailable. India's war with Pakistan in 1965 over Kashmir and in 1971 which resulted in the dismemberment of Pakistan and creation of Bangladesh are just two examples.

Raw considers Sindh as Pakistan's soft under-belly. It has, therefore, made it the prime target for sabotage and subversion. RAW has enrolled and extensive network of agents and anti-government elements, and is convinced that with a little push restless Sindh will revolt. Taking fullest advantage of the agitation in Sindh in 1983 and the ethnic riots, which have continued till today, RAW has deeply penetrated and cultivated dissidents and secessionists, thereby creating hard-liners unlikely to allow peace to return to Sindh. Raw is also involved similarly in Balochistan.

RAW is also being blamed for confusing the ground situation is Kashmir so as to keep the world attention away from the gross human rights violations by India in India occupied Kashmir. ISI being almost 20 years older than RAW and having acquired much higher standard of efficiency in its functioning, has become the prime target of RAW's designs, ISI is considered to be a stumbling block in RAW's operations, and has, therefore, been made a target of all kinds of massive misinformation and propaganda campaign. The tirade against ISI continues unabated. The idea is to keep ISI on the defensive by fictionalising and alleging its hand is supporting Kashmiri Mujahideen and Sikhs in Punjab. RAW'S fixation against ISI has taken the shape of ISI-phobia, as in India everyone traces down the origin of all happenings and shortcomings to the ISI. Be it an abduction at Banglaore or a student's kidnapping at Cochin, be it a bank robbery at Calcutta or a financial scandal in Bombay, be it a bomb blast at Bombay or Bangladesh, they find an ISI hand in it.

RAW over the years has admirably fulfilled its tasks of destabilising target states through unbridled export of terrorism. The India Doctrine spelt out a difficult and onerous role for RAW. It goes to its credit that it has accomplished its assigned objectives due to the endemic weakness in the state apparatus of those nations and failure of their leaders.

2

Indian Intelligence Bureau

The Intelligence Bureau (IB) is India's internalintelligence agency and reputedly the world's oldest intelligence agency. It was recast as the Central Intelligence Bureau in 1947 under the Ministry of Home Affairs. The reason for the perception may be because, in 1885, Major General Sir Charles MacGregor was appointed Quartermaster General and head of the Intelligence Department for the British Indian Army at Simla. The objective then was to monitor Russian troop deployments in Afghanistan, fearing a Russian invasion of British India through the North-West during the late 19th century.

In 1909, the Indian Political Intelligence Office was established in England in response to the development of Indian revolutionary activities, which came to be called the Indian Political Intelligence (IPI) from 1921. This was a state-run surveillance and monitoring agency. The IPI was run jointly by the India Office and the Government of India and reported jointly to the Secretary of the Public and Judicial Department of the India Office, and the Director of Intelligence Bureau (DIB) in India, and maintained close contact with Scotland Yard and MI5.

Dineshwar Sharma is the current director of IB, holding the position since January 1, 2015.

Responsibilities

Shrouded in secrecy, the IB is used to garner intelligence from within India and also execute counter-intelligence and counter-terrorism tasks. The Bureau comprises employees from law enforcement agencies, mostly from the Indian Police Service (IPS) or the Indian Revenue Service (IRS) and the military. However, the Director of Intelligence Bureau (DIB) has always been an IPS officer. In addition to domestic intelligence responsibilities, the IB is particularly tasked with intelligence collection in border areas, following the 1951 recommendations of the Himmatsinhji Committee (also known as the North and North-East Border Committee), a task entrusted to the military intelligence organisations prior to independence in 1947. All spheres of human activity within India and in the neighborhood are allocated to the charter of duties of the Intelligence Bureau. The IB was also tasked with other external intelligence responsibilities as of 1951 until 1968, when the Research and Analysis Wing was formed.

Activities

Understanding of the shadowy workings of the IB is largely speculative. Many a times even their own family members are unaware of their whereabouts. One known task of the IB is to clear licences to amateur radio enthusiasts. The IB also passes on intelligence between other Indian intelligence agencies and the police. The Bureau also grants the necessary security clearances to Indian diplomats and judges before they take the oath. On rare occasions, IB officers interact with the media during a crisis situation. The IB is also rumoured to intercept and open around 6,000 letters daily. It also has an email spying system similar to FBI's Carnivore system. The Bureau is also authorised to conduct wiretapping without a warrant.

Workings

The Class 1 (Gazetted) officers carry out coordination and higher-level management of the IB. Subsidiary Intelligence Bureaus (SIBs) are headed by officers of the rank of Joint Director or above, but smaller SIBs are also sometimes headed by Deputy Directors. The SIBs have their units at district headquarters headed by Deputy Central Intelligence Officers (DCIO)or ACIO (Assistant Central Intelligence Officer). The IB maintains a large number of field units and headquarters (which are under the control of Joint or Deputy Directors). It is through these offices and the intricate process of deputation that a very organic linkage between the state police agencies and the IB is maintained. In addition to these, at the national level the IB has several units (in some cases SIBs) to keep track of issues like terrorism, counter-intelligence, VIP security, threat assessment and sensitive areas (i.e. Jammu and Kashmir and such). IB officers (like their counterparts in R&AW) get monthly special pays and an extra one-month salary every year, as well as better promotions.

Constitutionality

IB was created on 23 December 1887, by the then British Secretary of State as a sub-sect of the Central Special Branch but there is no act of the Indian parliament nor executive order relating to the functioning of the IB. In 2012, a PIL was filed challenging the legality of IB.

Like other intelligence agencies, IB is also suffering from lack of parliamentary oversight and legal framework governing its functioning. A Private Bill titled Intelligence Services (Powers and Regulation) Bill, 2011 was introduced by then union minister Manish Tiwari.

However, it was withdrawn at the instance of then prime minister of India. Other suggestions have also been given in this regard but till the month of September 2014 there is no sign of any legal framework for IB. Intelligence infrastructure of India needs transparency and strengthening and reforms. A balance between national security requirements and civil liberties protection must also be maintained by India.

Ranks and insignia

Ranks of Gazetted Officers (Group 'A')

- Director Intelligence Bureau (post held by senior most Indian Police Service officer; and the only 4 starequivalent to Armed Forces Generals in terms of insignia. However it's a post and not a rank, and carries salary and status equivalent to a Secretary to the Government of India, or a 3 Star General.)
- Special Director (equivalent to Director General of Police)

- Additional. Director (equivalent to Additional Director General of Police)
- Joint Director (equivalent to Inspector General of Police)
- Deputy Director (equivalent to Deputy Inspector General of Police)
- Joint Deputy Director(equivalent to Senior Superintendent of Police)
- Assistant Director (equivalent to Superintendent of Police)
- Deputy Central Intelligence Officer

Ranks of officers in (Group'B')

- Assistant Central Intelligence Officer (equivalent Deputy Superintendent of police)

Ranks of Officers in (Group 'c')

- Junior Intelligence Officer- I (equivalent to Sub-Inspector of Police).
- Junior Intelligence Officer- II(equivalent to Assistant sub inspector)
- Security Assistant (equivalent to head constable)

There are different ranks of executives which adhere for managing and executing the goals of the organisation. Sometimes executives are compared with state police service ranks which are different from the ranks in IB.

Operations

The Intelligence Bureau reportedly has a lot of successes to its credit, but operations conducted by the IB are rarely declassified. Due to the extreme secrecy surrounding the agency, there is little concrete information available about it or its activities. The IB was trained by the Soviet KGB from the 1950s on wards until the collapse of the Soviet Union.

The IB was initially India's internal and external intelligence agency. Due to lapses on the part of the Intelligence Bureau to predict theSino-Indian War of 1962, and later on, intelligence failure in the India-Pakistan War in 1965, it was bifurcated in 1968 and entrusted with the task of internal intelligence only. The external intelligence branch was handed to the newly created

Research and Analysis Wing. The IB has had mixed success in counter-terrorism.

It was reported in 2008 that the IB had been successful in busting terror modules. It alerted the police before the Hyderabad blasts and gave repeated warnings of a possible attack on Mumbai through the sea before the November 2008 Mumbai attacks. On the whole, however, the IB came in for some sharp criticism by the media after the relentless wave of terror attacks in 2008. The government came close to sacking top intelligence officials soon after 26/11 attacks because of serious lapses that led to the 2008 Mumbai attacks. Heavy politics, under-funding and a shortage of professional field agents are the chief problems facing the agency. The overall strength of the agency is believed to be around 25,000, with 3500-odd field agents operating in the entire country. Of these, many are engaged in *political intelligence.*

THE INTELLIGENCE BUREAU

The reports of the Intelligence Bureau should facilitate, not hinder, people-oriented development; and strengthen, not erode, the democratic processes. For this to be possible, the report on any given subject should be well-researched, accurate and comprehensive. Then and only then can an intelligence report help the government and serve the public interest. The now-infamous IB report on non-governmental organisations fails to meet these criteria. The treatment of the Vedanta project in Odisha is an example of the one-sided treatment in the report.

The Intelligence Bureau's (IB) document criticising foreign-funded non-governmental organisations (NGOs), now available in the public domain, has provoked all-round protests from civil society. In a way, this controversy epitomises the recurring conflict between security considerations and civil liberties in democracies all over the world.

The well-known political scientist, Robert Jervis, summed up this dilemma by saying," Intelligence and intelligence services are simultaneously necessary for democracy and a threat to it". To balance security considerations with the imperatives of civil liberties is a formidable challenge, even in well-evolved democracies.

The IB is undoubtedly a premier intelligence agency, perhaps the world's oldest. Its roots are colonial, as is the case with many institutions in India. All such institutions view the rulers with awe and treat civil society as an adversary. IB's forerunner was the then British Indian Army's Intelligence Department set up in 1885 to gather intelligence on external threats against the British rule in India. In 1909, its successor, the Indian Political Intelligence (IPI) Office was established to keep a tab on Indian freedom fighters. It was recast in 1947 through an executive order, not a statute, as the Central Intelligence Bureau.

In the past, the IB's painstaking studies of emerging threat perceptions have significantly promoted national security. At the same time, during the last six decades of the IB's existence, there have been complaints that the ruling political executive at any given point of time relied heavily on the organisation to keep a tab on its political adversaries, a fact that should cause serious concern. On occasions, the agency's reports are known to have highlighted what is palatable to the ruling party and under-played what is not so palatable. Incomplete and inaccurate reporting could lead to biased decisions.

The IB's activities have remained insulated from public scrutiny. Instead of confining itself to gathering strictly security-related, factual information and reporting objectively to the government, the IB often ventured into areas of policy and ideology, alien to it. As in the case of most other wings of the government, attractive post-retirement assignments, including gubernatorial appointments, to those that headed the agency, eroded its functional autonomy. No wonder that some of these aspects have since come under judicial scrutiny.

What Is an NGO?

Coming to the role of the civil society entities, the term "NGO" in itself could mean a dedicated individual working selflessly among the people without any external support. It may be a group of like-minded persons working for a common cause, or a domestically funded body with its focus exclusively on a few well-defined social concerns, or a foreign-funded agency working in line with the donors' objectives. Even a political party is strictly speaking an NGO.

NGOs receiving foreign donations are required to register under the Foreign Contributions (Regulation) Act (FCRA), 2010, submit periodic reports on their activities to the Ministry of Home Affairs (MHA) and remain accountable to the government. Both FCRA and the Representation of the People Act (RPA) explicitly prohibit political parties or their candidates from accepting foreign donations.

In our country, feudal traditions are still dominant and political corruption rampant. As a result, the vast majority of the people in the low-income groups have remained far too diffident to be able to voice their concerns to the government. But, for the critical and exemplary role played by some outstanding individuals and NGOs over the years, their concerns would never have come to the forefront of political discourse, and the government would not have found it necessary to make appropriate policy interventions. For example, it was at the instance of some NGOs that the Supreme Court had examined the ambit of Article 19 of the Constitution and provided a landmark interpretation, leading to the enactment of the Right to Information Act (RTIA) 2005, and the mandatory requirement that candidates contesting elections to public offices must disclose their antecedents upfront. Both these measures have far-reaching implications for governance.

However, over the years, the character of the NGOs has undergone a marked change globally, due to ever-increasing infusion of funds from large corporate entities and philanthropic agencies. The Center for Civil Society Studies at Johns Hopkins University suggests that even after excluding the funds provided for religious purposes, the so-called non-profit sector is now a $1.1 trillion per annum industry, employing well over 19 million fully paid workers, equivalent to the world's eighth largest economy! Several Indian NGOs have benefited from this huge influx of funds. The donors' objectives in such cases often came into conflict with the policies of the ruling political executive, either for valid or for not so valid reasons. Several corporates have tried to leverage their donations to choke competition and create favourable markets for themselves. Some have even tried to influence government policies. For example, if there are genuine NGOs promoting small, distributed renewable electricity generation units on the one side, there are an equal number of dubious corporate-funded NGOs on the other side, promoting large centralised renewable energy units to develop a market for the corporates. It would, however, be grossly unjust to say that all foreign-funded NGOs in India are engaged in objectionable activities.

POLITICAL PARTIES AND FOREIGN DONATIONS

The contagion of foreign-corporate-donations is not just confined to the NGOs alone. It has since spread to several political parties in India. In this connection, the Delhi High Court, in its order dated 28 June 2014, directed the government to take appropriate action. As already pointed out, unlike the NGOs duly registered under the FCRA, monitored by the MHA, political parties are prohibited from accepting foreign donations, both under FCRA and RPA. Foreign companies donate to political parties, not out of magnanimity, but in return, for huge quid pro quos, such as, getting the regulatory authorities to acquiesce to statutory and human rights violations committed by the corporates, the government allowing them to indiscriminately exploit scarce natural resources, and so on. Ironically, one is yet to come across an IB report on such illegalities and the havoc that the political donations have resulted in. Before pointing fingers at foreign-funded NGOs, political parties and intelligence agencies should therefore introspect on this. The intention here is not to discount the value of the intelligence reports but to help IB reorient its own role in line with the public interest.

Intelligence reports will subserve the public interest better if they alert the government to statutory violations, whether committed by the corporates, the NGOs or the political parties. Their reports should facilitate, not hinder, people-oriented development; and strengthen, not erode, the democratic processes. For this to be possible, the report on any given subject should be well-researched, accurate and comprehensive. Then and only then can an intelligence report help the government and serve the public interest. How far does the IB report meet these requirements?

Vedanta Project

In the very first paragraph of its report, the IB refers to the case of Vedanta's bauxite mining project in Odisha as an example of how NGOs have "stalled development projects" by encouraging people's agitations against them. One would then expect the IB to support such a strong assertion by highlighting all the facts that are relevant to the project and explaining the concerns of all its stakeholders so that the government can take a balanced view on the various issues.

In all, 15 Indian NGOs, supported by three foreign agencies, namely, Amnesty International, Action Aid and Survival International, have encouraged people's resistance and successfully stalled the Vedanta project. Amnesty International received 1,25,000 pounds sterling from George Soros Open Society Foundation in 2009. The IB's report cited the industrialist, Sajjan Jindal's interview on 12 September 2013 to *The Economic Times* on inter-corporate rivalry and Jindal confirming that, the corporates routed Rs 50 crore annually to Indian NGOs to encourage agitations to their advantage. The report then tries to quantify the direct and the indirect fund flows from several corporates to NGOs through the other two foreign agencies, namely, Action Aid and Survival International. Finally, in Para 44 of the report, there is a fleeting reference to the main stakeholders, namely, "Dongria Kondh tribes of Nyamgiri Hills" in Rayagada and Kalahandi districts, where Vedanta's bauxite mining is to be undertaken without any mention of the rights of the tribals and their travails. In particular, explicit on any FCRA violation by the NGOs.

Ignoring Tribal Rights

Vedanta's mining project is located in the upper reaches of the Niyamgiri Hills, the habitat of the Dongaria and Kutia Kondhs (primitive tribal groups), lying within an area notified under the Fifth Schedule to the Constitution. The local tribals not only enjoy certain land and forest rights under the Constitution, but also the rights available under the Panchayats (Extension to Scheduled Areas) Act, 1996 (PESA) and Scheduled Tribes and Other Traditional Forest Dwellers (Recognition of Forest Rights) Act,

2006 (FRA). The state government and the Ministry of Environment and Forests (MoEF) initially ignored these rights and routinely granted clearances to Vedanta.

When the government remained insensitive to the concerns of the tribals, the local NGOs took up their cause, leading to public protests against mining. Heeding the tribals' protests and realising that the clearances so far given were deficient, the MoEF appointed a four-member expert committee, under the chairmanship of N C Saxena, a highly regarded former senior civil servant, to review the clearances. The Saxena Committee visited the area, interacted with the government and the tribals and concluded that mining in the area would not only violate the constitutional and customary rights of the tribals, but also the various environment and forest laws, as well as PESA and FRA.

On 24 August 2011, the MoEF revoked clearances for the project. On a writ petition filed by the Orissa Mining Corporation against the MoEF's order, the apex court considered the matter in detail and upheld the order. The Supreme Court, inter alia, observed, the Saxena Committee's evidence as reviewed by the FAC (Forest Advisory Committee) and read by me as well is compelling. The violations of the various legislations, especially the Forest (Conservation) Act, 1980, the Environment (Protection) Act, 1986, and the Scheduled Tribes and Traditional Forest Dwellers (Recognition of Forest Rights) Act, 2006, appear to be too egregious to be glossed over.

The apex court ordered the state government to seek the local tribal gram sabhas' consent for the project, as required under PESA and FRA, before proceeding further. Accordingly, gram sabha meetings were held and the tribals, in one voice, resolved not to have the project anywhere near their habitat.

It is surprising that the IB did not refer to any of these facts, especially the statutory violations and the landmark judgment pronounced by the apex court.

The IB report is equally silent on the huge contributions made by Vedanta to the major political parties in India. A cursory look at the company's annual reports, corroborated by the contribution reports submitted by political parties to the Election Commission of India (ECI) would have helped the IB make its report more meaningful, cogent and relevant.

A quick look at the NGO's website would have revealed that a highly regarded former comptroller and auditor general of India is a sitting member of its international board. Apparently, the IB was unaware of this. While the IB report has emphasised the concerns against the NGOs as perceived by the corporates and others, the report would have been more balanced had the IB taken pains to point out the tribals' point of view, the nexus between the corporates and the political parties, the quid pro quos that hurt the public interest and the judicial scrutiny that confirmed the statutory violations.

One could go on evaluating the present IB study with reference to POSCO and other cases mentioned in it. The picture that emerges is similar, one of deficient and incomplete analysis.

In the very first paragraph of its report, the IB summarily estimates the negative economic impact of the NGO activities to be 2%-3% of gross domestic product per year. In his insightful article, Surjit Bhalla (2014) refutedthis questionable estimate on the basis of facts. He concluded by saying that "law-abiding NGOs, foreign or domestic, are an asset to any society because they enlarge the debate via research and advocacy".

Lessons from the IB Report

From the controversy arising from the IB report, there are important lessons to learn, for the government, for the intelligence agencies, the NGOs and the political parties.

(1) The government should enlarge the democratic space for disadvantaged sections to enable them to articulate their concerns better and participate in decision-making.

(2) Instead of treating civil society as an adversary, the intelligence agencies should appreciate its genuine concerns and report them to the government. There is need for public accountability through parliamentary oversight of these agencies. World over, such reforms are taking place. The Justice and Security Act 2013 enacted by the United Kingdom to bring its intelligence agencies under

parliamentary oversight is the most recent example of this.

(3) The leaked IB report has indicted several genuine individuals and NGOs whose contribution to the welfare of the people has been immense and whose credentials are impeccable. To tarnish their names thus is patently unjust. Such indictments tend to suppress voices of dissent which form the essence of any democracy. The IB should introspect on this and take corrective action immediately.

(4) Like other public institutions, NGOs too should subject themselves to greater public scrutiny and comply with the law of the land. In a country like ours, where millions of people belong to the low-income groups, NGOs that splurge funds and indulge in profligacy are bound to be questioned.

(5) Instead of driving the NGOs to rely on foreign funding, the government should perhaps institute an independently administered fund to support their efforts, without any strings attached. Genuine NGOs have a crucial role to play in a democracy.

(6) Political parties too should agree to come within the purview of the RTI Act and promote transparency in the way they function. By reducing their own dependence on external funding, they could set an example to the civil society to do likewise.

The sooner the urgency of such reforms is realised, the better it would be for the well-being of the country.

3

Indian Defence Intelligence Agency

The Defence Intelligence Agency (D.I.A.) is an organisation responsible for providing and coordinating intelligence for the Indian armed forces. It was created in March 2002 and is administered within the Ministry of Defence.

It is the nodal agency for all defence related intelligence, thus distinguishing it from the R&AW. Much of the agency's budget and operations are classified. DIA has control of Indian Army's prized technical intelligence assets – the Directorate of Signals Intelligence and the Defence Image Processing and Analysis Centre (DIPAC). While the Signals Directorate is responsible for acquiring and decrypting enemy communications, the DIPAC controls India's satellite-based image acquisition capabilities. The DIA also controls the Defence Information Warfare Agency (DIWA) which handles all elements of the information warfare repertoire, including psychological operations, cyber-war, electronic intercepts and the monitoring of sound waves. Its operations are highly classified and has several success to its credit which will remain a secret.

Together with Research and Analysis Wing (R&AW) & Intelligence Bureau (IB), DIA forms a part of India's nodal intelligence setup.

The Defence Intelligence Agency (DIA) was created in March 2002. The new agency combines the intelligence networks of all three armed services of India (Army, Navy, and Air Force) and is the product of long-standing demands by the armed forces for such an organization and the recommendation of the Group of Ministers's report investigating the intelligence failures leading to the Kargil incident of 1999.

According to Defence Ministry officials, the DIA, unlike the previous Directorate of Military Intelligence, is to benefit from ample resources and manpower to allow it to fulfil its role.

The agency's primary task it to track troop movement in countries neighbouring India. The agency, unlike the Directorate General of Military Intelligence, is also assigned the mission of monitoring terrorist groups operating both within and outside of the country and is to have a section dedicated to gathering intelligence on terrorist groups and monitoring the internal security threat.

The agency's mission is to be accomplished using satellite and high-altitude aerial reconnaissance imagery.

The creation of the DIA significantly reduces the reliance of the Indian Armed Forces on civilian intelligence agencies such as IB and RAW for information and enemy troop assessments, as was the case during the conflicts of 1948 with Pakistan, 1962 with China, 1965 and 1971 with Pakistan and during the Kargil intrustion of 1999. During these operations, intelligence products were often found to contain discrepancies.

Prior to the creation of the Defence Intelligence Agency, the Armed Forces' military intelligence capability was limited to a few Field Intelligence Units (FIU) tasked only with gathering intelligence during ongoing operations and with maintaining liasion with foreign military forces.

The head of the Agency is the Director General; the first of which was Lt Gen Kamal Davar, an armoured corps officer and former Director General of Mechanized Forces. The Director General reports to the Defence Minister and, should the position be created, to the Chief of Defence Staff. The post is to rotate among all three of the armed services.

The Director General of the Agency is assisted by the Deputy Director General. The first to hold that position is Air Marshal S C Malhan.

Creation

The creation of an intelligence agency coordinating the intelligence arms of the three military services had long been called for by senior Indian military officers. It was formally recommended by the Cabinet Group of Ministers, headed by the then Deputy Prime Minister of India Lal Krishna Advani. The Group of Ministers investigated intelligence lapses that occurred during the Kargil War and recommended a comprehensive reform of Indian intelligence agencies. The Defence Intelligence Agency was created and became operational in March, 2002. As part of expanding bilateral cooperation on gathering intelligence and fighting terrorism, the United States military also provided advice to Indian military officers on the creation of the DIA.

Organisation

The Director General of the Defence Intelligence Agency is the head of the entire organisation. The director general is the principal advisor on intelligence to the Minister of Defence and the Chief of Defence Staff. The post of the director general will be held in rotation between the three armed services. The first Director General of the DIA was Lt. Gen. Kamal Davar, former director of the mechanised forces of theIndian Army. The Director General of the Agency is assisted by the Deputy Director General, held first by Air Marshal S. C. Malhan. The DIA will also be involved in intelligence support groups with the IB and RAW to coordinate military and civilian intelligence agencies and create an intelligence community.

Functions

Before the creation of the Defence Intelligence Agency, the military intelligence capability of India's armed forces was limited to Field Intelligence Units (FIU), the Directorate of Military Intelligence and separate intelligence arms of the services. These distinct services were not able to effectively coordinate intelligence operations and sharing of information. The armed forces also heavily relied on civilian intelligence agencies such as the Research and Analysis Wing and the Intelligence Bureau. With wide-ranging resources and functions, the DIA will be superior to and coordinate the Directorate of Military Intelligence, Directorate of Air Intelligence and theDirectorate of Naval Intelligence.

THE DEFENSE INTELLIGENCE AGENCY

The early weeks of this month have seen the formation of India's Defense Intelligence Agency (DIA). Lt. Gen. Kamal Davar was appointed as its first chief. Much of the public debate on this development has swung between extreme viewpoints and focused intensely on the publicly stated fissures in the intelligence system. This chapter will attempt to examine some of the issues confronting this newly formed agency in a more secular light. The 'Intelligence Process' involves seven basic steps: collection, analysis, production (assessment), dissemination, coordination, action, and evaluation. A sufficiently advanced intelligence handling system divides responsibility for these parts to independent groups of people, thereby maximizing oversight and minimizing overlap. In India we are still in the process of building such a system. Collection, analysis, and action are relatively easier steps to carry out. Production, dissemination, coordination and evaluation however are immensely complicated. If one carefully reads writings about intelligence issues in India, the difficulties experienced in these steps become apparent.

The bulk of intelligence work in India is carried out by the Research and Analysis Wing (R&AW) and the Intelligence Bureau (IB). The various services intelligence directorates namely the Directorate of Military Intelligence (DMI), the Directorate of Air Intelligence (DAI), Directorate of Naval Intelligence (DNI) and some other agencies are involved also but their activity is smaller by comparison. The R&AW and IB agencies are composed largely of civilians. Military personnel are often deputed to these agencies, but the letter of the law and concerns of deniability limit the use of serving military officers in some types of activity (especially collection and action). Democratic countries look down upon excessive involvement of military personnel in intelligence activity. The role of serving officers on deputation is therefore limited mostly to helping with 'production'. In cases where greater involvement of Military personnel becomes necessary, they are asked to resign from the forces and re-hired as civilians. As the Defense Intelligence Agency will be comprised largely of serving officers, the 'rules' will limit its role in physical collection of secret intelligence. The DIA may however choose to expand the existing Open Source Intelligence (OSINT) collection capability. Similar 'rules' apply to several 'action' related issues. It appears at first glance that the DIA personnel will be involved only in creating a support structure for various 'actions'.

In India repeated complaints have surfaced from various sources over the quality of intelligence production. As the military is one of the biggest consumers of the intelligence product, its complaints have been the most visible. It is impossible to dismiss all complaints as being parochial. A Defense Intelligence Agency does offer certain advantages in terms of quality of production. These are stated below:

a) The composition of the Indian military reflects the immensely diverse Indian society. Only a member of the military with years of service can truly comprehend the complex social divisions and hierarchy in the armed forces. Consequently this detailed knowledge should in theory enable the member to better mould the intelligence product to suit the needs of others in the military.

b) Understanding military technology requires considerable specialization. Such specialization exists in the military. With the growing use of technology in warfare there is a greater need for specialist participation in higher-level assessment.

c) Post-Chagai there is an emphasis on speed and reliability in all parts of the intelligence chain. At lot of the activity in this area will require the kind of discipline largely available in military service. Though in India, the entire responsibility for this targeting cannot be given to the DIA, its inputs will be critical.

In order to avoid becoming narrow and insular in its approach, it is strongly advisable that the DIA hire civilians to assist with the production process as well. Presently the country has a large surplus of trained professionals, a good number of universities and research centres. With suitable encouragement at least some of these could be induced to help with the task at hand. The author feels its it also a good idea to routinely farm out small production tasks to private think tanks and strategic forecasting companies. Such a measure will do a lot to improve the standard of public debate on security and strategic issues in India. The task of 'dissemination' of intelligence offers many challenges. The DIA should hopefully attempt to address some like:

a) The battlefield awareness of the Indian soldier could do with some improvement. Modern technology has improved the ability to reach the soldiers in the battlefield, but knowing exactly what information to give to them is important. This is in part a production issue as well, however any steps taken to improve the ability of the lowest rungs of the military to comprehend intelligence briefings is most welcome.
b) Dissemination of information in the field has its vulnerabilities; modern technology does not overcome all of them. Here too the exact way in which the information travels from source to destination will need much more thinking. There are SIGSEC and COMSEC capabilities but more agile management is always desirable.
c) The sharing of information once it is collected with other agencies is a contentious issue especially in areas of 'concurrent responsibility'. This was the *Achilles heel* of the old system. Whether modern technology can address some of this only time will tell.

'Coordination' at the DIA (local) level will involve managing the exact division of responsibilities between the various service intelligence directorates. Given the diversity of the theatres of operation, it will be a challenge to integrate the service intelligence directorates while maintaining an emphasis on corroboration. The DIA will also act as the liaison between the MOD sub-units and other intelligence centres. At a higher level, the DIA will have to brief the CDS and, and 'coordinate' its production activity with other agencies under the supervision of the Intelligence Coordination Group. This part may take time to settle down.

This brings us to the last and most important step of 'evaluation'. This step deals with ensuring accountability within the intelligence handling system. Without this critical step, the entire process of intelligence can become skewed. Indian intelligence agencies have suffered a fair bit because of a lack of transparency in their evaluation procedures. Most of these are cloaked in the same veil of secrecy that surrounds other operations. The result is that any public motifs about the agency subsequently color perceptions about accountability. There is a great deal to be learnt from the mistakes made by other agencies in the past. Undoubtedly the new system of intelligence management involving the Intelligence Coordination Group etc... will have evaluation methods built into it, but a mechanism for auditing and reviewing performance within the DIA itself is highly advisable.

In conclusion, it is important to bear in mind that none of the problems have evaporated merely with the formation of the DIA and 'old enemies' may reappear even in the new setup. However in the past the formation of such agencies was overshadowed by severe economic crises. The crises skewed the development of these agencies. The DIA has been born at a relatively calmer time. This should have a very positive impact on the DIA. While all the items above seem within reach, it is important to realize that all of them are big challenges. It is also vital to bear in mind that the DIA is not some magic wand capable of conjuring instant geopolitical victory. The DIA will eventually be a sharp tool, most effective in skilled hands. The formation of the DIA is many years overdue and Lt. Gen. Kamal Davar has an enormous task ahead of him, so with that in mind, the author congratulates him and wishes him and the staff of the DIA the very best of luck on their endeavour.

4

Role of R&AW in Liberation of Bangladesh

Covert action capability is an indispensable tool for any State having external adversaries. Its purpose is not just collection of intelligence, but the protection of national interests and the safeguarding of national security through deniable actions of a political, economic, para-diplomatic or para-military nature. A State resorts to covert action if it finds that its national interests cannot be protected or its national security cannot be safeguarded through conventional political, economic, diplomatic or military means or if it concludes that such conventional means are not feasible.

Any intelligence agency worth its salt will have a covert action capability ready for use, when necessary. The Governments of some countries openly admit the availability of such a capability in their intelligence agencies, but not the details of their operations, which have to be secret and deniable. Others don't admit even its existence.

In India too, the IB, under the foresighted leadership of the late B.N.Mullik, its second Director, had a limited covert action capability for possible use. The covert action division of the IB played a notable role in the then East Pakistan to counter the activities of the ISI in India's North-East.

In India, one tends to think that Pakistan's use of terrorism against India started in 1989 in Jammu and Kashmir (J&K). It is not so. It started in 1956 in Nagaland. The ISI trained the followers of Phizo, the Naga hostile leader, in training camps set up in the Chittagong Hill Tracts (CHT) of East Pakistan. It also provided them with safe sanctuaries in the CHT from which they could operate in the Indian territory through northern Myanmar.

In the 1960s, it started providing similar assistance and sanctuaries to the Mizo National Front (MNF) headed by Laldenga in the CHT. The ISI's set-up in East Pakistan also enabled the Naga and Mizo hostiles to establish contact with the Chinese intelligence. This paved the way for the training of the Naga and Mizo hostiles in training camps set up by the Chinese intelligence in the Yunnan province of China.

It was partly to put an end to the activities of the ISI in India's North-East from East Pakistan that Indira Gandhi decided to assist the Bengali-speaking people of East Pakistan in their efforts to separate from Pakistan and achieve an independent State to be called Bangladesh. This was in the wake of the widespread disturbances in East Pakistan in the beginning of 1971 following the refusal of the military regime of Pakistan headed by Gen. Mohammad Yahya Khan to honour the results of the December,1970,general elections in which the Awami League of Sheikh Mujibur Rahman won a majority in the Pakistani National Assembly.

When the people of East Pakistan rose in revolt in March,1971, the R&AW was two and a half years old. It was still in the process of finding its feet as a full-fledged external intelligence agency, with a hardcore of professional intelligence officers capable of operating under cover in foreign territory as well as across the border in the neighbouring countries.

The R&AW had inherited from the IB its intelligence collection and covert action capabilities relating to Pakistan and China. These were not up to the standards of the intelligence agencies of the Western countries and Israel. They had many inadequacies, which had become evident during the Chinese invasion of India in 1962, during the Indo-Pakistan war of 1965 and during the counter-insurgency operations in the North-East.

The late Rameshwar Nath Kao, who headed the external intelligence division of the IB, was appointed by Indira Gandhi as the head of the R&AW when it was formed on September 21,1968. In the first few months after its formation, he gave it two priority tasks— to strengthen its capability for the collection of intelligence about Pakistan and China and for covert action in East Pakistan.

A little over two years is too short a time to build up an effective covert action capability, but the R&AW managed to do so. It went into action the moment Indira Gandhi took the decision to help the people of East Pakistan achieve their independence from Pakistan.

The 1971 war against Pakistan was not a war won by India alone. It was a war jointly won by India and the people of East Pakistan. It would be wrong to project that India was the architect of an independent Bangladesh. India's role was more as a facilitator than as a creator.

Without the desire and the will of the people of East Pakistan to be independent, there would have been no Bangladesh. Their sacrifices for their cause were immense. How many of them were brutally killed by the Pakistan Army! How many of the Bengali intellectuals were massacred by the Pakistan Army and by terrorist organizations such as Al Badr and Al Shams created by the ISI! It is their sacrifice, which laid the foundation for an independent Bangladesh. What India did under the leadership of Indira Gandhi was to make sure that their sacrifices were not in vain.

The Indian Armed Forces under the leadership of Field-Marshal (then General) S.H.F.J. Manekshaw and the Border Security Force (BSF) headed by the late K.F.Rustomji overtly and the R&AW and the IB covertly ensured this. But, they would not have been able to succeed as well as they did without the political leadership provided by Indira Gandhi and the phenomenal work done by the civilian officials of West Bengal, Assam and Tripura in organizing humanitarian relief for the millions of refugees who crossed over into India from East Pakistan.

Indira Gandhi's dramatic decision to ban all Pakistani flights over India to East Pakistan in retaliation for the hijacking of an

Indian Airlines flight by two members of the Jammu and Kashmir Liberation Front (JKLF) to Lahore in January,1971, paved the way for the ultimate victory in East Pakistan. When the Pakistani aircraft tried to fly round India over the sea by availing of re-fuelling facilities in Sri Lanka, Indira Gandhi pressurized the Government of Sri Lanka to stop providing the re-fuelling facilities. This greatly weakened the ability of the headquarters of the Pakistani Armed Forces in West Pakistan to send reinforcements to East Pakistan and to keep their garrisons in East Pakistan supplied.

The R&AW's role was five-fold: Provision of intelligence to the policy-makers and the armed forces; to train the Bengali freedom fighters in clandestine training camps; to network with Bengali public servants from East Pakistan posted in West Pakistan and in Pakistan's diplomatic missions abroad and persuade them to co-operate with the freedom-fighters and to help in the freedom struggle by providing intelligence; to mount a special operation in the CHT against the sanctuaries and training camps of the Naga and Mizo hostiles;and to organize a psychological warfare (PSYWAR) campaign against the Pakistani rulers by disseminating reports about the massacres of the Bengalis in East Pakistan and the exodus of refugees.

The flow of intelligence to the policy-makers from the R&AW and the IB was continuous and voluminous.

This was facilitated by the co-operation of many Bengali public servants of East Pakistan and by the poor communications security of the Pakistani Armed Forces.

One of the first acts of Kao after the coming into being of the R&AW was to set up a Monitoring Division headed by a distinguished retired officer of the Army Signal Corps to collect technical intelligence (TECHINT) from Pakistan and China and a Cryptography Division, headed by a cryptography expert from the IB. While the performance of the Monitoring and Cryptography Divisions in respect of China was unsatisfactory, they did excellent work in intercepting electronic communications within West Pakistan as well as between West and East Pakistan and in repeatedly breaking the codes used by the Pakistani authorities for their communications.

ON LIBERATION STRUGGLE OF BANGLADESH

The struggle for liberation going on in Bangladesh at present is very important, particularly for India, because our country, India, surrounds that landmass within which this freedom movement is going on. So, even if other countries of the world may consider questions like the past and future of that country, the consequence of the present political movement there, and what would be the character and form of the government there to be an exclusive affair of that country, the understanding of the Indian people should not be like this.

Our relationship with the people of Bangladesh is a historical one. We have naturally a very close relationship with them, an organic link so to say, and we share a cultural linkage between us. In fact, once we all were people of the same country, although now we are divided not only into two states but also into two nations, into two countries.

On this, I feel strongly that an important point should be made clear to those people in our country who are dreaming of an immediate unification of East Bengal and West Bengal through this freedom movement in Bangladesh and are giving such a call. That is, although we were once people of the same country, but that feeling of one nationhood does not exist in reality today in either of the two countries. Evidently, centring round the two countries two kinds of national pride and nationhood have developed independently of each other. It is true that there is alikeness in the sense of Bengali entity between us, the people of West Bengal, a constituent state of India and the people of Bangladesh. But other than cultural and linguistic similarity between the two, the national feelings or nationalism of theirs is now completely different and separate from ours. We cannot afford to forget this any way. Whatever the importance of their independence movement to us, we should bear in mind while extending support to this freedom movement that Bangladesh aspires to exist separately along with its own sovereignty and freedom. So, we have to extend support to their freedom movement with the attitude of helping in the development of their sovereignty and national life.

If our help and support to them be motivated by such an attitude which shows we are displaying sorts of undesirable superintendence over them or that we are one and the same as they are and that we shall now reunite to be one again, then our help will convey a different message to them. Such an attitude of ours will not go down well with the very mentality that has prompted the national uprising in Bangladesh. Our party considers that if this type of attitude is expressed now it will create harmful repercussions instead of helping the freedom movement of Bangladesh. By this, it is the forces opposing the freedom movement there that will be strengthened. As a result, the very purpose of our extending sympathy and support towards the freedom movement will be adversely affected and the freedom movement will, in reality, be weakened in a way.

In order to correctly grasp the nature of the new, independent national awakening that we are now witnessing in Bangladesh, it is necessary to discuss briefly the history of its development. After two states were created by artificially dividing India, the Indian nation, by force and Pakistan was constituted as a separate state, in the course of time there was an attempt to develop a Pakistani nationalism, and it was developing too. But due to a number of reasons a common Pakistani national feeling and a common mental make-up comprising West and East Pakistan could not develop. The first reason being the great distance that separates West Pakistan from East Pakistan (East Bengal), and so geographically they are completely separated from each other. Secondly, it turned out to be a difficult task to build up a unity in the cultural sense, too, between these two ends of Pakistan, because from the point of view of language and culture, too, one has no similarity with the other. Besides, for advancement, industrialization and economic development of West Pakistan, the ruling elite of Pakistan had virtually reduced East Bengal into a colony. Despite the population of East Pakistan being more than half of the population of Pakistan, one fourth of the total revenue generated from East Bengal used to be spent for East Bengal. And the rest was used to be utilized to increase the wealth of West Pakistan. So, just as the imperialists try to develop and advance their own countries by making use of their colonies as the sources of raw materials and keep them subjugated, so also even after the formation of Pakistan as one country and one state the ruling elite of West Pakistan turned East Bengal into such a colony.

But the realization of this colonial exploitation and resentment against it did not grow among the people of East Bengal at the initial stage and the sense of Pakistani identity did not disappear from the minds of the people of East Bengal just in a day. The people of East Bengal used to consider themselves to be Pakistani for a long period. But the public opinion in East Bengal was developing gradually against the ruling elite of West Pakistan centring round language, culture and on question of economic development, and various demands began to surface. But the ruling elite of West Pakistan could make out neither the latent meaning nor the historical direction of these movements. On the one hand, they sought to suppress these movements with brutal state power and, on the other, by fomenting religious fanaticism they tried to artificially create and sustain the concept of one single, undivided Pakistani nationalism among the common people. But they had forgotten that a nation does not develop only on the basis of religion.

Nation is a product of history. In the course of history, from the strong urge of development of capitalism an economic centralization and an awareness of one centralized identity develops within a given geographical boundary, and on its basis, keeping language, culture and all-round economic development in the forefront, a nation is born as a product of a historical process. This is the history of formation of all modern nations. Viewed from this angle, if Pakistan had not been carved out from India by forcible imposition of an unjust will, the present two countries – Pakistan and India – would now have existed together historically as a single nation. But India, which developed as a nation in the course of history, encountered vivisection because of political bungling of the ruling class, giving birth to two nations, two countries, two patriotic sentiments. But even after Pakistan was constituted as a separate state, East and West Pakistan could not give birth to a single patriotic feeling. Though the ruling class of Pakistan tried utmost to build up one national mental make-up in the geographically separated two parts of Pakistan basing upon religion in the name of Islamic state completely negating language and culture, it is however a fact that Pakistani nationalism arose in the form of two divided streams of nationalism. The national mental make-up in West Pakistan developed in one way, and the national mental make-up of East Pakistan, which is now being called Bangladesh, began to develop in a separate way. This Bangladeshi mental make-up which developed separately within the geographical environment of East Bengal is a new nationalist feeling or nationalism. But along with this, it must never be forgotten that this nationalism of Bangladesh is an offshoot of the Pakistani nationalism itself.

If the different provinces of West Pakistan are taken separately, then it will indeed be seen that Pakistani nationalism does not exist there, too, in one integrated form, because powerful resentment of the people against oppressive rule of the autocratic ruling clique, strong resentment among the other nationalities against the Punjabi Muslim nationality's attempt of domination are very much extant. Even so, because of the geographical contiguity between the provinces of West Pakistan, the resentment among the different nationalities there could not take on such a widespread and intense form so as to give rise to an independent national feeling as in Bangladesh. But because of lack of this geographical contiguity between the two parts of Pakistan, thereby giving birth to two different geographical environments within the same state of Pakistan, and because of the presence of burning resentment against colonial exploitation, slavery and political subjugation of the oppressed people of East Bengal coupled with their wounded feeling, new nationalist feelings and a new mental make-up started to develop there gradually. As a result though the ruling clique of Pakistan could maintain unity to a certain extent among the different nationalities of West Pakistan, but another current of the same Pakistani nationalism has found separate expression in the form of the nationalism of East Bengal, i.e. East Pakistan.

This national awakening of Bangladesh by freeing itself from the religious superstition of Pakistani nationalism has given birth to a completely new nationalism, and through a truly national awakening and upsurge of patriotic awakening and in the form of patriotic upsurge, Bangladesh has appeared in the pages of history as a new nation. It can no longer be called a part of Pakistan. To the people of Bangladesh the rulers of West Pakistan have assumed the character of imperialist rulers, and for Pakistan East Bengal has become a separate nation.

This sense of separate national identity, patriotism of East Bengal is completely different. If the Pakistani ruling clique tries to suppress it by force or are even successful to suppress this national uprising by any means – even then, it should be borne in mind that a new national consciousness has awakened in East Bengal.

Bangladesh will rise up again and again demanding a separate state and would ultimately succeed in establishing an independent state. The freedom movement of no country can be ultimately kept suppressed in the old way. But at the same time it has to be borne in mind that because the national uprising that has occurred in East Bengal has been born as an offshoot of Pakistani nationalism, so within the present national uprising a compassionate and affectionate feeling is still at work as regards Pakistan in the thinking of the people of East Bengal even today.

So, those who are giving a call for unification of two Bengals, making freedom movement in Bangladesh a basis of their argument or dream about uniting the two Bengals, considering it from the point of view of India's special interest, should remember that the state of mind of the common people of Bangladesh is not like that and they too are not viewing such an eventuality in this way. It is true that we were once united and that we shall again be united in future.

But not in this way. If unification is attempted this way right now, the very cause of freedom movement of Bangladesh will suffer most. Those who are dreaming of uniting two Bengals here and now should always remember that even if a socialist state structure comes into existence in both these countries – India and Bangladesh – under the leadership of the working class in future, even then unification of the two countries cannot be accomplished immediately by abolishing the common boundary of the two countries. The national boundary of two neighbouring countries cannot be abolished immediately after the establishment of socialism. The right to do away with the national sovereignty and national frontiers does not come about with the establishment of the working class state. It has to be kept in mind that even after the socialist revolution is victorious, because the form of the socialist revolution from country to country is still national, although the content of those revolutions are socialist and international, it is not possible for any socialist country to do away with the national boundaries, unless and until this national form of socialist revolution is historically exhausted. If socialist countries try to achieve unification by forcibly doing away with the national boundaries, mutual bickering will develop among the socialist countries themselves. Because, the problems of dominating over and imposing one's view on the other will appear, and what is more, the sense of national pride, which is mixed as an alloy in socialist revolution, being hurt because of the attempt of such forcible union will give birth to the mentality of going out of socialism itself. As a result it is the socialist revolution that will suffer setback.

United we shall all be in the distant future, when the entire human society, all the countries will merge into one human society. Revolution will be victorious throughout the world, one human civilization will arise combining all the countries on the globe. Prior to that, our duty is to selflessly extend help for the advancement and development of a nation, once it emerges through a struggle. The very attitude of getting united here and now has no need at all. To the people of both the countries such an attitude at this juncture is harmful indeed. So my request to those who are helping the freedom struggle of Bangladesh is that they should do so in the spirit of a volunteer.

However, this does not mean that no interest of ours is at all involved in this freedom struggle of Bangladesh. Our only interest is, if a revolutionary government is established in Bangladesh through the freedom struggle then that will constitute an effective conducive force, no doubt, helping the revolutionary movement of India. At least we should hope that such an independent sovereign state will be established in Bangladesh as would not act as a puppet state of imperialism, a state which would defend the very spirit of independence and extend support to the greater cause of anti-imperialist freedom movements all over the world, which would extend support and backing to the democratic movements in the country and help the country to advance gradually on the path of progress and socialism. So we want this. If a government like that of Tunku Abdur Rahman is established in Bangladesh after having achieved independence then we shall surely be unhappy. At the same time it is very important to bear in mind that there is no question of withdrawing our wholehearted support to the cause of independence of Bangladesh, fearing that a government similar to that one led by Tunku Abdur Rahman is going to be installed there. Because, under what type of leadership a government will be formed there is a problem that concerns the freedom-fighting people of that country. If the people of Bangladesh install a moderate leadership in power through the present freedom movement then any attempt by us or any external forces to unsettle that would jeopardize the entire independence itself. Regarding the character of the leadership, there can be discussion and exchange of opinion. But whatever may be its character, if there is no disagreement on the question that the present movement in Bangladesh is a freedom movement, then all honest, democratic-minded patriotic people should extend support to this freedom movement.

What type of government they can form right now and what character that independent state would assume ultimately will totally depend on the character of the leadership in their movement and on the level of political consciousness of the people in this mass movement, in this freedom movement. Besides, which of the different political forces who are unitedly fighting side by side in the present freedom struggle of Bangladesh, despite differences of opinion among themselves, has more influence over the people and a stronger organization among them, and particularly which of them would be able to provide leadership on the present armed struggle, would also largely determine the character of the future government there. Because, the freedom struggle of Bangladesh having passed through the different stages of democratic mass movements, such as peaceful movements, rallies and processions and strikes, has at present taken the form of an armed mass uprising against the armed brutal and barbarous Pakistani military administration, and the people there are engaged in life and death struggle for independence. So, among those who are together conducting the freedom struggle, it is the party which in course of providing consistent leadership will be able to really build up and organize armed struggle, grasping its intricacies, with the utmost dedication and ability that will ultimately come into the leadership of the freedom movement. Today whoever may be in the leadership there in the freedom movement, finally it is the party that is able to advance the armed struggle successfully that is bound to come into the leadership and into state power after independence.

Here we want to tell those who are engaged in the freedom struggle in Bangladesh a very important point. We are observing there a simmering discontent against the leadership of Mujibur Rahaman among a section of the participants in the present armed freedom struggle. In our country too we are observing that a section of those who claim to be revolutionaries are even saying such things that Mujibur Rahaman is an agent of foreign powers. Perhaps they are saying so by observing the connections Mujibur Rahaman has with the heads of different capitalist states and the friendly relationship of Yahya with China or, perhaps they have doubt as to whether Mujibur Rahaman will fight with the ideal of socialism; so they are propagating these. But they are not considering at all if the present freedom movement being conducted under the leadership of Mujibur Rahaman is called a conspiracy of the imperialists because he has connections with different capitalist-imperialist states, then this would be tantamount to opposing the very freedom movement itself. It is true, from whatever information we have received so far about Mujibur Rahaman, he appears to us also to be a moderate leader. But this notwithstanding, it has to be borne in mind that the people of entire Bangladesh have plunged into such a gigantic struggle for complete independence centring round the leadership of Mujibur Rahaman, and it is Mujibur Rahaman who stands at present as the focal point of national unity and solidarity among the common people of Bangladesh in the present national uprising there. Hence, whoever may try to discredit the leadership of Mujibur Rahaman at present by raising various questions would not only be isolated from the public mind but would even weaken the solidarity that is there in the freedom movement. And doing this will in reality be tantamount to ultimately strengthening the hands of the barbarous military clique of Pakistan. So, we strongly feel that whatever may be the internal contradiction or differences of opinion about the political character of the leadership, those points should never be brought to the forefront now and must not, on any account, relegate to the background the cardinal question of the unity of the people that has emerged in the freedom movement of Bangladesh.

We consider that one vital point needs to be raised now with emphasis by us and on behalf of the freedom fighters too. It is this: Bangladesh must have full independence. There can be no settlement by moving away even an inch from this. Because, we are observing that a conspiracy is on in this regard on the part of the US and other interested circles. They perceived that if attainment of independence of Bangladesh cannot be frustrated despite all their efforts, then it would have to be supported ultimately. But if Bangladesh can be tied up with Pakistan before that through a confederation by UNO mediation then that would be the best thing from their point of view of protecting their interest. Therefore, the US official lobby is at present very active in this regard. To hatch such a formula, many a seasoned head is kept busy in the capitals of the various capitalist countries. On all these there is need for remaining very much alert.

Hence, today it is necessary for the freedom fighters of Bangladesh to raise this slogan with the greatest emphasis that there must be no yielding even by an inch from full independence, that is, full sovereign national independence. If any leader, any section or any party participating in the freedom movement raises the question of discussing any terms short of full independence – then whoever they might be, or whatever might be the leadership, they would have to be gradually isolated from the freedom movement. Whoever extend their help in the present struggle for achieving full independence would have to be treated as friends of the freedom movement whatever might be their political philosophy, concept or ideology. If it is felt that Mujibur Rahaman would frustrate the freedom itself if the freedom movement is conducted under his leadership, then he would have to be got committed to full independence. A national liberation front would have to be raised in Bangladesh at present along with whoever desire to fight on the demand of full freedom. And in this freedom struggle the unity of this national liberation front would have to be protected like the apple of the eye. It is true that at the initial stage Mujibur Rahaman sought merely the right of autonomy in Bangladesh and tried for that. But later on he himself demanded full independence of Bangladesh and sought to provide leadership to the freedom struggle. So, the greatest need now is to carry forward the freedom movement of Bangladesh through forming a national liberation front combining with him, although his is a moderate leadership. Our appeal to the freedom fighters of Bangladesh is that they should not accentuate the differences among themselves centring round the question of leadership at the present time. Because, by that they would for all practical purpose weaken the very freedom movement.

Then again, the revolutionary forces there should keep in mind one thing in this context. That is, if they desire to provide leadership to the freedom struggle simply by levelling criticism against Mujibur Rahaman they cannot do so. For, revolutionary leadership cannot be established through mutual bickering and recriminations. There is only one way to establish this leadership. If they can correctly conduct ideological struggle within the national liberation front that would be formed in the present liberation struggle while maintaining unity of the front and, if they have the ability to provide the best form of leadership in the political, organizational and military spheres in the armed struggle that would perhaps continue there for a long time, then the leadership of the national liberation front will surely ultimately pass on to their hands. No one can stall it. And if Mujibur Rahaman himself provides that leadership then no one has any reason to object to it. Everybody will have to obey that leadership. One cannot go against the freedom movement just because, say, a feudal lord is providing the leadership. Those who do so frustrate the main objective. Therefore, the freedom fighters of Bangladesh would have to remember that by raising slogans against Mujibur Rahaman while scrambling for leadership, it is the national uprising that has developed at present centring round Mujibur Rahaman against the misrule and exploitation of Pakistan that would be weakened and divided, and the hands of the Pakistani military ruling clique would be strengthened thereby.

We think it necessary to have some discussion on the role of China on the Bangladesh issue. Because, thanks to the bourgeois newspapers of our country confusions of various types have appeared not only among a section of the people but even in the parties who call themselves Marxist-Leninists. In our country while, on the one hand, the circles known to be anti-China reactionary forces are conducting various kinds of misleading propaganda about the role of the Chinese government on the Bangladesh issue centring round the friendly relation between China and Pakistan, and, on the other hand, some Naxalites in our country who swear by revolution, on seeing the friendly relation between China and Pakistan and the silence that China has maintained so far on the Bangladesh issue, are opposing the freedom movement of Bangladesh and are branding this whole movement as a conspiracy of imperialism and the Indian government.

We think that such approach by China is being taken mainly with the objective of making use of the internal contradictions of the world imperialist-capitalist camp in the interest of revolution as a tactical means for weakening the imperialist camp ultimately. Many perhaps think that this policy of China is being pursued by its narrow nationalist interest. Be that as it may, I do not intend to enter into the controversy as regards different perceptions and interpretations regarding this policy of China. I want to say only this much that it is because of the sustained endeavour and successful diplomatic efforts of the Chinese government for a long period that the USA has not been able to get Pakistan committed in its favour on Vietnam despite Pakistan being a member of SEATO and its being almost totally dependent on the USA economically and militarily. It is the successful diplomatic efforts of China that pulled Pakistan to the opposite direction. However much reactionary the present military ruling clique of Pakistan may be, China does not want to mar in any way its friendly relations with Pakistan for the present in order to correctly handle the contradictions in the imperialist camp and to manoeuvre Pakistan to keep it as far as practicable out of American influence. Again it can in no way be concluded from this that just because it has to maintain this relationship, China does not support the national liberation struggle of Bangladesh or it will not support it in future if the necessity arises.

We are placing before you briefly how we have understood the question of China's not openly supporting the freedom movement of Bangladesh. China is observing that because the leadership of this movement has been in the hands of moderates like Awami League, Mujibur Rahaman from the very beginning, no preparation for building up armed struggle by organizing the masses or armed resistance on a broad scale could develop there from the outset. So, the entire freedom movement under the leadership of Mujibur Rahaman is still mainly in an agitational form. The possibility of developing effectively and widely armed struggle or armed resistance or a national liberation army from within is remote in the near future. But the geographical and logistic advantages are not such that China can effectively extend any sort of help in developing armed struggle or building up national liberation army within Bangladesh by providing direct aid. So, China perceives that to extend open support to the freedom movement at this hour of the movement will in reality be reduced to a verbal support only, which will effectively mean nothing at all so far as the freedom of Bangladesh is concerned except extending just a moral support. On the contrary, there is the possibility of some adverse reactions following from it. One: if China openly supports the freedom movement at the present moment, then the support of the official and unofficial circles of different capitalist countries, including the USA and Britain for the freedom struggle under the leadership of Mujibur Rahaman and Awami League that is there, the support of the Soviet Union to it that is there now behind it, and the zeal and initiative with which the Government of India is now supporting the freedom struggle of Bangladesh — there is every possibility of all these to be disrupted. Two: moreover, on the pretext of the involvement of China in the affairs of Bangladesh the USA could directly intervene in Bangladesh by taking the side of Pakistan, as it did in Vietnam. Three: China knows it very well that if they openly support the freedom struggle, Pakistan would not accept it. Moreover, as a result of such an action on the part of China, the pressure that is being exerted in this matter by some capitalist countries upon Pakistan today, this would be to a certain extent hindered, and what is more, the possibility of achieving freedom by Bangladesh, even under a moderate leadership like that of Mujibur through the mediation of countries like the USA, UK and USSR would also encounter diplomatic hurdles. So China is perhaps thinking that so long as the concrete condition is not created from within for conducting a protracted armed struggle inside Bangladesh itself, any open utterance in support of the freedom struggle of Bangladesh would do more harm than good at the present hour. Four: on the other hand, Pakistan, which China has been able to isolate at least to a certain extent from the imperialist camp through prolonged successful diplomatic effort, would again be pushed completely into the arms of imperialism.

Perhaps by considering these aspects, China is of the opinion that it is not proper to say anything openly in support of the freedom struggle of Bangladesh at the present moment. So, it would not be proper to conclude just now from China's saying nothing openly in support of the freedom struggle of Bangladesh that China does not support this freedom struggle or that it is opposing it. Sometimes back China sent a note to the Indian government protesting about the malicious campaign against China that the anti-Chinese reactionary forces of our country with the patronage of the Government of India had been conducting by using as an opportunity the well considered silence of China on the freedom struggle of Bangladesh, on the one hand, and China's friendly relation with Pakistan, on the other. If the language of this protest note is carefully studied that will provide an inkling of the Chinese attitude on the freedom struggle of Bangladesh despite its present silence. At one place of this protest note China has said,

"India is slandering against us that we are supporting the Pakistan regime against the freedom fighters of East Pakistan." The use of the word "freedom fighters" in this note is significant. If it is read carefully this should not escape anybody's attention.

The seasoned diplomats of different capitalist countries and the bourgeois newspapers have correctly understood that this action of desisting right now from openly supporting the freedom struggle of Bangladesh and the efforts by them to maintain a friendly relation with Pakistan in various ways do in no way mean opposing the freedom movement. And as because they too have realized it they are showing over-enthusiasm over Bangladesh and are frantically trying to find out a kind of settlement on the Bangladesh issue as quickly as possible. Because they know if the freedom struggle runs along the path of a protracted armed battle, then in the near future there would be emergence of revolutionary leadership in the freedom movement by ousting the moderate leadership and it would ultimately come under the influence of China. And because they understand this, they are so concerned. Otherwise, that the imperialist-capitalist powers are opposing liberation struggle everywhere is known to all.

But the different capitalist countries are supporting the freedom fighters of Bangladesh and trying their utmost to bring about a settlement of the issue. This can have only one object; it is this that centring round Bangladesh nationalism, once the movement to attain complete independence has begun it cannot be ultimately quelled at the present times. From many a bitter experience, they have understood this. So long as the leadership of this movement remains in the hands of a moderate like Mujibur Rahaman and his Awami League, if the independence movement could be brought to an end within that time by hastening a political settlement through transfer of power, then even after achievement of independence would it be possible for the capitalist imperialist countries to maintain relation with it in their own interest, and maybe it would also be possible to keep Bangladesh loosely connected with Pakistan. So, clearly it is the apprehension that the entire freedom movement may ultimately come under the influence of China which is at work behind all these so-called pious attempts of theirs. It is this line of thinking that has been at work behind all the activities of the rulers of different capitalist countries and the propaganda publicity being given by the bourgeois press as is becoming evident from some facts, information and opinion being expressed by foreign papers and magazines, which I am going to refer to in the hope that you will certainly understand from it how they are viewing the role of China on this issue.

The Guardian which represents British imperialist interest has advised Pakistan like this: Even if it be possible for you to suppress this present uprising, still you must not forget that this movement will raise its head again and again, so it would be a prudent act to concede to the demand for freedom and confer independence now while there is still time. That means, they want to say, it is necessary to politically settle the issue with the elected representatives of the Awami League while there is still time, before the Chinese influence increases. The idea that they hold is more or less like this: Pakistan has foolishly created the present situation. Had they conceded to Mujibur's six point demand at the very outset there would have been no ground for the emergence of a situation as it is obtaining today. But when there has been an eruption, there is no escape from the situation. Now if Pakistan again behaves obdurately, then not only independence would have to be conceded, there will be no kind of relationship between Bangladesh and Pakistan. Perhaps another Vietnam would be created and the whole of Bangladesh would come under the influence of China. So this cannot be in any way to the liking of any capitalist country, including the USA.

It is in their own self-interest that such bourgeois newspapers are trying to correctly grasp the role of China on this issue. Addressing Pakistan, The Guardian, a trusted representative of British imperialism, further told: You say that China is your friend, but if your generals are surprised to find the freedom fighters fighting with Chinese automatic rifles tomorrow and start pondering how the freedom fighters have procured those Chinese automatic rifles despite China being a friend, then we can't help calling them 'idiot generals'. A newspaper of Paris, too, has tried to impress upon Pakistan almost in the same vein: You think that as China is your friend so China will stand behind you against the freedom fighters of Bangladesh, evidently, you have not at all gauged the politics that China is pursuing. The same newspaper has remarked in a jocular vein in one place: China is assuring you about Bangladesh that 'Don't worry, we are behind you, if you are endangered by anybody we shall see to it. If you who belong to the Pakistani ruling circle place faith in such kind of assurances, then your condition too will be much like the proverbial patient'. The story which they have mentioned runs like this – when a patient suffering from heart trouble went to a doctor, the doctor was assuring him: "Don't worry, I shall cure your liver." This newspaper of Paris has tried to warn Pakistani ruling clique: the assurances given by China to you about Bangladesh are much like the assurance given by that doctor.

In The Statesman, too, a foreign correspondent has recently penned a feature titled, Chinese Riddle. There, too, he has tried to bring home the same point in a roundabout way. Although a bourgeois journalist he could by no means subscribe to the view that despite various Chinese assurances to Yahya, the present role of China on the question of Bangladesh is one of opposing the freedom struggle. Rather, he is of the opinion that if the freedom struggle there takes the form of a protracted armed struggle, then it is China that will help Bangladesh most. The gentleman has advised Pakistan to be on guard for that while there is still time.

Jayaprakash Narayan, the Sarvodaya leader, by apprehending that the Chinese hand is behind the freedom struggle of Bangladesh, has also tried to convince the state heads of different countries of the world that unless they are able to effect an immediate settlement of the Bangladesh issue by exerting pressure on Pakistan, there is every possibility of Bangladesh falling under the complete influence of China in the near future.

So, it is seen, that starting from the British, French and American newspapers and journals, even Soviet Union all are trying to impress upon Pakistan the same point. Everyone's apprehension is: there is every possibility that the whole freedom movement may ultimately come under the influence of China. Notwithstanding all these developments what is very striking is that the so-called revolutionaries and Marxist-Leninists of our country are not in a position at all to realize this. On the other hand, the Indian government which is indulging in progressive verbiage these days and spreading verbal fireworks in support of the freedom fighters of Bangladesh – they too have either not been able to grasp this, or else, despite being fully aware of the Chinese attitude to the freedom struggle are conducting motivated propaganda against China to thwart the increasing Chinese influence on the movement.

In this context my appeal to the Government of India: if they are really desirous of providing effective and selfless support and help to the freedom fighters of Bangladesh then it is China even more than the Soviet Union that can be the only reliable force for India in this regard. Hence, rebuilding friendly relationship with China should be considered as a most urgent task for effectively supporting the freedom struggle of Bangladesh and for preserving the independence of Bangladesh even after attainment of freedom – leaving aside the question of India's own progress and development.

So frankly speaking, the type of propaganda against China going on in our country, that China is on the side of Pakistan against the freedom struggle – I find this difficult to believe.

We think it necessary to discuss a little about the role of the Indian government on the question of the present freedom struggle in Bangladesh. For, whatever support we may have as the people of India for this freedom struggle of Bangladesh, how much help can we provide compared to their present need without any effective role of the Indian government? What we can do is, if the freedom fighters of Bangladesh so want and if the Government of India allows us to go, then we can send thousands of volunteers there. They can cross the border and can fight and can mobilize resources, collecting and handing over funds and medicines. In the liberation struggle along with the freedom fighters, they can lay down their lives. But the help that is most needed today by the freedom fighters of Bangladesh is the effective help for building up an invincible liberation army with innumerable armed detachments, equipped as guerilla fighters. The people of our country cannot effectively provide what is necessary for these. That can only be provided by the Indian state, the Indian government.

So, we as the people of India have to be vigilant so that in the matter of providing help and assistance to Bangladesh there is no perfunctory attitude of the government to provide some help anyhow. It can never be allowed. The type of help which the freedom fighters of Bangladesh need most will have to be provided if it is needed. Risk, to a certain extent, we must take. But I am not in favour of taking risk in a hot-headed manner. My personal opinion is that now we can take risk to a great extent. If we do not extend help along wrong lines, if we do not do that suffering from any misconception or with any ulterior motive, then to my mind in the present situation we can take a lot of risk. Apprehensions are being expressed from some quarters that in an eventuality China may join hands with Pakistan against us, but such talks are evidently inconsequential. That China is really against this movement and is on the side of Pakistan – I don't believe this at all. It is impossible for me to believe this. My thinking on this is not like that of the other parties. I have found by minutely observing that the language of China's utterances on this issue — though of a different type from our language — but I cannot say that China is speaking in a totally unfamiliar language. I have already made some discussion about this and highlighted certain pertinent facts and information.

If we want to make the Government of India extend exactly that kind of help which is necessary to effectively help the freedom fighters of Bangladesh then first of all the democratic movement of our country needs to be established on a firm basis centring round a definite ideology and principle. I firmly hold this to be the most urgent primary task in the matter of extending help to Banlgadesh. If we can do this, then that would be a worthwhile job. The government would do its bit. Yes, we are seeing, to facilitate that no one seems to be lagging behind in running and rushing from here to there. But if our democratic movement is not on a correct footing, then our government may do many things in the name of extending help, which are not in conformity with the desire of the Indian people. Since our government is the embodiment of the interests of the monopolists, it is not the way we want to help the people of Bangladesh but instead, naturally, there is every possibility that the very expansionist desire of the monopoly class will manifest itself through its policies, and in extending help to the struggle of Bangladesh this expansionist motive is likely to work. So, the sole guarantee of getting the Government of India to extend the kind of selfless help to the cause of the people of Bangladesh in the way we want it, lies in raising the democratic movement to a higher conscious level. If the minimum standard of democratic movement is not adequately present in the country, then in the name of supporting and helping the movement of Bangladesh the Government of India may naturally try to exercise its big power-like superintendence or try to spread its sphere of influence there to a certain extent in the interest of trade and commerce and may even try to install a puppet government in Bangladesh. In order to bring a restraining effect on such motives working behind the government help and if it is to be ensured that governmental help moves along the right direction, the unity of the left and the democratic forces has to be restored again within our country, consciousness has to be developed within the movement regarding these motives behind the government help and above all a minimum code of conduct and minimum ideological cohesion has to be developed within the democratic movement. Because although Indian capitalism is backward compared to that of the western capitalist countries, it has already been transformed into monopoly capitalism and has itself acquired imperialist character. This can be proved easily with the help of statistics. It may not have that wide dimension like the imperialist design of the western countries, but I do not accept that Indian monopoly capitalism does not have a limited expansionist desire, as just the opposite of this is being indicated by data, statistics and also theoretically.

So, we think when we are exerting pressure on the Government of India to the effect that something must be done immediately : "Do something and do it courageously! Extend help in as many ways as possible and as required !" – then, at the same time, the left and democratic forces of our country would have to be constantly vigilant as to whether the government is doing this in consonance with the aspirations of the people of our country. And the greatest guarantee for ensuring this, as I have already said, lies in unity, a minimum ideological standard in the democratic movement and a firm policy which are altogether lacking in our country today. So, it is seen that even in the matter of helping Bangladesh, the strengthening of our own democratic movement, conducting it along the right direction is of crucial importance.

But you see the Government of India has said many things expressing its sympathy towards the freedom struggle of Bangladesh. But under different pleas it has not yet accorded recognition to the independent government that the freedom fighters have formed in the meantime. The Government of India's sole concern in according this recognition seems to be to see whether USA, UK, USSR or China do so or not. I like to ask: what prompts you, the Indian government, to hesitate so much? As far as other states are concerned, which way to see this freedom struggle – that is their concern.

It is a fact that India has certainly recognized this freedom struggle. Then why is it failing in according recognition to this independent government of Bangladesh? Giving recognition to this independent government of Bangladesh and extending every help in equipping its liberation army as an effective armed resistance force had been the bounden duty of the Indian state and government. Yet the Government of India is cleverly trying to avoid that responsibility.

They are no doubt providing some money and medicines in various ways as help along with sweet words, but they seem to be afraid of openly according recognition to the independent government there and of providing arms on a large scale and if necessary military help. It seems from the attitude of the Government of India that it is perhaps thinking, if the freedom fighters are defeated and if the ruling military clique of Pakistan is able to suppress this freedom movement then its diplomatic relation with Pakistan will deteriorate. In that case the question that arises is: in extending all-out support to the independence struggle should the success-failure prospect be considered to be the determining factor by the Indian government? If that be the case then what is the necessity for them to mouth high sounding words in support of the freedom struggle? By this the Indian government is simply showing hypocrisy, resorting simply to trickery, trying simply to keep the freedom fighters under its fold. The intention is clear; if the freedom fighters are victorious then it would show off how much help India had provided to them. And if the freedom fighters are defeated, then the relation with the state of Pakistan would also remain intact; the diplomatic relation with it would not be snapped. That is the trickery. So our party considers that on the matter of extending effective help to the freedom struggle of Bangladesh no amount of appeal to the Indian government will yield any result. For that due pressure is to be exerted on the Government of India by organizing intense movement, and the people should therefore step forward in that direction.

We too, as the people of India, have many things to do in the matter of helping the freedom struggle of Bangladesh right now. We should first of all speedily create an organized volunteer corps so that we can effectively help in this freedom struggle. Those who want to work as volunteers in the freedom struggle of Bangladesh should enroll their names en masse as volunteers. Our party has already appealed to the people to form a volunteer corps. If we can build up a powerful volunteer corps then we can effectively demand of the Government of India to transform that volunteer force, ready to participate in the freedom struggle, into an armed force and to make arrangements for sending them to Bangladesh. But if this volunteer corps is not equipped, then as some unorganized people crowding there and making a fuss we shall only create hindrance to the freedom fighters but not render any effective help to them. So the building up of this volunteer corps and making it ready in every respect is a compulsory task for us. Besides, we need to send food, clothing and medicines to help them. But along with this we have to see that this help does not fall into the hands of bad elements. Because, various types of forces are at work there. It will have to be ensured that our help properly reaches the organizations of those who are the real freedom fighters and those organization and parties which are constituents of the freedom struggle. So we would like to request those non-government organizations who are collecting money and materials in aid of the freedom struggle of Bangladesh to send the aid preferably through a political party or through the government of our country in case they have no faith in political parties.

Lastly, we consider it necessary to examine to a certain extent the present situation of the democratic movement of our country in the context of the liberation struggle that is going on at present in Bangladesh. I think that the people participating in the democratic movement of our country and the political parties here have much to learn from this freedom struggle of Bangladesh. Because, whereas the people as well as different progressive democratic parties there are carrying on unitedly a life and death struggle against the military ruling clique, the mass movement of our country is today split because of internal squabbles. Deception has taken root within the democratic movement itself of our country. Confusion abounds in it at all levels. Moral standard, political principles, ideological standard within this democratic movement are collapsing. It is true that we conduct struggles here on various democratic demands, but we the participants forget that we have a duty and responsibility to the society too along with that struggle. The ethical standard that we reflect in all these democratic movements is that if any one is friendly to me or subscribes to my opinion, we applaud him like anything. But even if someone is a leader and fighter in the democratic movement we do not feel any qualms in abusing him in filthy language if we have even the slightest difference of opinion with him. The political principles and ethical standard within our democratic movement has come down to this level today.

In this context we wish to remind you again one thing. We discussed this point in my speeches in many a previous meeting. It is that there were many flaws and shortcomings in the freedom movement that we conducted against British imperialism in this country. We are aware of that. During the freedom movement we failed to conduct ceaseless struggle against religious bigotry and blindness in order to develop democratic values and mindset within the society. A national feeling and patriotism free from blind religious superstitions did not guide the independence movement of our country. Rather things like religious revivalism, religious reformist outlook were mingled with the freedom movement of our country. As a result, we conducted the national freedom movement of our country on the basis of a religion-oriented nationalism. We could not hold aloft the flag of cultural and social revolution within the freedom movement. Yet even at that time we observed that before the freedom movement, because it was progressive for the country despite all its shortcomings, there was a clear-cut political ideology. A certain standard of morality and ethics guided the youth and the people participating in the mass movement then. The sons and daughters of the country fought against British rule but did not turn indisciplined. They sacrificed their lives in the face of bullets but did not indulge in misbehaving with anybody.

But today we see that the strength of the so-called communists, socialists, of those parties who claim to be big revolutionary parties has increased. Their influence over the youth and the people of the country has increased. They themselves claim this too. But we know that a revolutionary ideology by its very nature reflects higher ethical and cultural standard. So, if their ideology is really a revolutionary one, then the people who are awakened by its influence should have reflected a higher cultural and ethical standard through its magic touch. Then how to explain this degradation of cultural and ethical standard among them? Why this indiscipline and aimlessness among them? Why is such aimless reckless mentality engulfing the youth under their influence? It is true that they fight on democratic demands and raise slogans on democratic rights too, but do not feel any sense of duty and responsibility to the society. Why is such a mentality working at different levels of the democratic movements in our country? The more the strength of such so-called revolutionaries is increasing, the strength of their parties increasing, the more are indiscipline, misbehaviour, vainglory and arrogance among the youth increasing, and the ethical standard is sliding downward. Then does this not undoubtedly prove that their strength is increasing indeed, but that strength is an evil force, it does not signify increase of strength of a democratic force or real revolutionary force. Because, had there been an increase in strength of the revolutionary force then at least a new life throb, vitality should have been generated among the people who are moving under their influence. They should have been vibrant with new principle, morality and ideology. They would have laid down their lives, but nobody could have said they were indisciplined and indulging in misbehaviour.

Till now the people who are now fighting for freedom in Bangladesh are not vocal about Marxism-Leninism, the noble revolutionary ideology. Again it may be that there are such talks within, but to us these are inaudible. Their present struggle is one for attaining national independence, which we had conducted from 1922 to 1947. There they are fighting against the enemy, making supreme sacrifices, embracing death. They themselves are dying and are killing the enemies too. But they are restrained, they are not oppressors. But in the democratic movement of our country in case of difference of opinion we abjure ethics and culture, we lack sobriety. When differences of opinion with others crop up we do not hesitate to insult people, harass people, thrash them — friend and foe alike – and behave in a most uncivilized manner.

What is the basic cause for the difference between the movement in Bangladesh and the movements in our country — I feel that also needs to be considered. It should be remembered that the creed of nationalism which was avowed as noble ideal during the freedom movement in our country – that same creed of nationalism is at work in the present freedom struggle of Bangladesh. But after acquiring political independence and establishment of a bourgeois state in our country that creed of nationalism exercises no longer any efficacious role in our social life today. It can no more imbue the people of our country. But, what the big parties known as Marxist-Leninist are propagating in the name of ideology is not Marxism-Leninism, it is only slogan mongering. We know that the kernel, the essence of any ideal, any philosophy, any noble ideology is ingrained in its moral standard, in the cultural and ethical standard. Marxism is a noble revolutionary ideology. Consequently the essence, the kernel of this noblest revolutionary ideology too lies ingrained in its cultural and ethical standard. Hence, if the ethical standard of what the so-called big communist parties of our country tout as Marxist ideology has descended to a level that is lower than the standard of those who are fighting for freedom of Bangladesh, then the pertinent question is: where will the mass movement conducted by them lead our country to? When the leaders and cadres of big political parties seek to justify the correctness of their political line citing instances of their widening mass support, I would like to remind them one or two incidents of history. At times it may so happen that thousands of people rally under the slogan of the pseudo-revolutionaries, even the reactionaries. The people get confused, the whole country is emotionally surcharged – such incidents occurred many times in history. Once a man like Hitler had got the people of entire Germany, from the working class to the intelligentsia, frenzied with the slogan of progress, the 'dawn of a new era'. But this sea of humanity which once thronged behind Hitler – does it prove that Hitler rallied the masses behind the cause of Germany's real development, advancement, socialism and progress? Or is it not the truth that Hitler confused the people of Germany with slogans of progress, even if temporarily? At that time keeping in mind the low standard of public consciousness Hitler very deceitfully raised some such slogans in Germany to which the entire people of Germany gravitated by thinking that to be the only course for advancement of Germany, for progress of the working class, having failed to understand the implications of the slogans. As a result, catastrophe befell the whole of Germany.

What was the condition of the communist party of Indonesia? Before the uprising of reactionary forces in a small country like Indonesia, almost the size of Bangladesh, the total number of members alone of the communist party there was three million. After inducting their own people to the armed forces, police and at different levels of government administration, the communist party there was thinking that they had almost completed the task of revolution. What did they actually do? They played with revolution. Those who want to accomplish revolution by aligning with the capitalist or imperialist state structure and by using its administrative machinery, what they do in the name of revolution is to expose the entire party to the exploiting state machinery. Such an inconceivable incident happened in Indonesia. So when the tide of reaction came, the military turned against them. Then the entire party got smashed up in the face of the onslaught of state power. Because, as a result of exposing the party to the state nothing of the party was hidden from the state machine. This was the major flaw, speaking in the technical sense, or in other words the manner in which they wanted to accomplish revolution — from that point of view the main failing.

Their second failing was that along with the increase in strength of the party, they did not gauge at what level lay the cultural and ethical standard of the people, how much was their sense of responsibility and duty to the society, and they did not make continuous efforts to uplift the same. In our country too when the workers of the so-called big revolutionary parties parade on streets, shout slogans and make various gesticulations in a highly objectionable way and resort to obscene talks and use of abusive language in the event of some arguments or disagreements, it at once reveals the level of the ethical standard of this democratic force. If the ethical-cultural standard of people and the party workers remains at this level, then it should be remembered that those workers and masses who at one time install these so-called revolutionary parties in power by raising slogans in elections and at another time bring them down, they may any day, coming in contact with another political current, become a powerful weapon in the hands of reaction, as happened in Indonesia.

These so-called big revolutionary parties are not even for once thinking of taking lesson from history, how the hundreds of thousands of people who used to assemble at the call of the Indonesian communist party got lost at an extremely adverse situation faced by the party. They were the multitude of people who had stood behind the Indonesian communist party in meetings, rallies and processions carrying banners and raising slogans. But when the storm of reaction broke out, when the state power combining with the religious fanatics went out to annihilate the party, it was found that the larger section of the population of the whole country stood behind the state power against the three million communists. The communist party there never bothered to gauge the mindset, the direction of the thinking of that larger section of the population outside their three million members. Because, they became visibly smug with the strength of three millions. They never cared that however large their party membership might be, it was a minority compared to the vast masses of the whole country.

This is why all the great revolutionaries taught that a revolutionary party must always keep in view that compared to the greater mass of the people of the country those who actively participate in the struggle with the revolutionary party in a most effective manner during revolutionary movement, constitute a minority.

While conducting the revolutionary movement no revolutionary party can afford to disregard the mental orientation and direction of thinking of this vast mass of the people. It is true that as because this greater mass of the people is not organized, it shrinks in fear in the face of rallies of hundreds of thousands, it cannot build up any effective resistance even if it wishes to. But it is also true that unless the mass movement is conducted in the correct fashion and if shortcomings and flaws persist within the democratic movement, then this mass of people develop resentment in them against it. If a party is not able to gauge the mentality of the people in time and does not correct its policy, conduct and method of work accordingly, then unknown to it this resentment among the masses gradually goes on spreading out, and some day at an opportune hour the masses throw out that party in the face of a violent storm.

It is this danger that has appeared today in the mass movement of our country, too. So we need to view the freedom struggle of Bangladesh in a greater perspective. We have to take lessons from the freedom struggle of Bangladesh in order to free our minds, free ourselves from obliviousness of our great heritage, from self-conceit, and get rid of cultural degradation and for advancing the standard of the mass movement of our country. If we can do this, then that will help in a great way in properly conducting the democratic movement of our country in the coming days.

RAW AND BANGLADESH

RAW AND BANGLADESH is an illuminating book written by Mr. Zainal Abedin, a senior Journalist of Bangladesh and published by Fatema Shahad in November 1995. It is printed by Madina Printers and distributed by Madina Publications, 38 Banglabazar Dhaka - 1100, Bangladesh.

Mr. Zainal Abedin is a former student leader and a freedom fighter who crossed over to India in 1971 for military training, joined the Mujib Bahini and fought for the freedom of Bangladesh alongside the Indian Army. Mr Abedin holds a masters degree and a BEd.

In the book Mr Abedin ponders over the events of 1971 and 'one particular incident chastises me the most', he writes, It was in April 1971 in the district headquarters of Noakhali, his home district, where the freedom fighters were gathering, that he met his neighbour. A middle-aged, mature and sober person, his most staunch supporter and admirer. 'He got hold of my hand and took me to a nearby restaurant'. Narrates Mr Abedin, 'As we sipped tea, he started narrating the history of pre-1947 Bengal. He mentioned gory details of how the Hindus used to treat the Muslims and explained the reasons for creation of Pakistan. He said that struggling for one's right is different from break up of the country. He stated that India would first weaken us by breaking our unity and then exploit us. He added that his life time experiences had told him that India would never be sincere to Muslims. With tears in his eyes and hands trembling with emotions he quipped, 'Are you again going to make us the slaves of the Hindus?'.

Mr Abedin admits that he did not give any serious thought to the urging of his neighbour at that time as he was too young and emotional. 'The said question now haunts me often', he writes. 'The realization of what lay in store for us started soon after I crossed over to India. The attitude of our Indian handlers and trainers indicated that they treated us (the Freedom Fighters) not as friends but as agents. The real Indian face lay bare after the surrender of Pakistani forces, when I saw the large scale loot and plunder by the Indian Army personnel. The soldiers swooped on everything they found and carried them away to India. Curfew was imposed on our towns, industrial bases, ports, cantonments, commercial centres and even residential areas to make the looting easier. They lifted everything from ceiling fans to military equipment, utensils to water taps. Thousands of Army vehicles were used to carry looted goods to India. History has recorded few such cruel and heinous plunders. Such a large scale plunder could not have been possible without connivance of higher Indian authorities'.

From the conduct of the Indian Army it was evident that they treated Bangladesh as a colony. That was the time the question asked by his neighbour seriously surfaced in Mr Abedin's mind. He feels that 'India through her notorious deeds has proved time and again that she is not our friend but an arch roguish foe', which has posed a grave threat to the independence and sovereignty of Bangladesh.

'It is now evident', writes Mr Abedin, 'that India helped the creation of Bangladesh with the aim that it would be a step forward towards the reunification of India'. Soon after the creation of Bangladesh, India let loose all forces at her command to cripple the newly born country. Their aim was to precipitate its collapse and eventual merger with India to realize part of the Brahmanic dream about 'Akhand Bharat'. The most significant player of this heinous game is India's notorious intelligence agency, the Research and Analysis Wing commonly known as RAW. Established in 1968 and still 'eating into the vitals of Bangladesh'.

The author explains the main sources of inspiration for India's intelligence agencies and her foreign policy - the ancient pundi-cum-political, Koutillaya, Prime Minister of Chandra Gupta. In his book 'Arthshastra' written about 300 BC Koutillaya laid down three guidelines: when your country is weak pursue the policy of peace; when your country becomes militarily strong follow the policy of war; when another state seeks your help, apply double standards. In his book Koutillaya has recommended six principles as the basis for foreign policy. These deal with peace, war, neutrality, military preparedness, formation of alliance and duel policy. Koutillaya's final dictum is. 'Power is the ultimate truth and the main aim of an organised state should be to obtain power'.

The author shows how Mrs Indira Gandhi was an ardent follower of Koutillaya and subscribed fully to his policy of waging 'battles of intrigues' and 'secret wars' to achieve her unholy objectives. She created RAW, a secret intelligence agency functioning directly under the Prime Minister to pursue her ambitious but nefarious agenda particularly in the South Asian region.

The Research and Analysis Wing (RAW), explains the author, was raised in September 1968 from the foreign Intelligence Desk of Intelligence Bureau (IB). Mr Rameshwar Nath Kao was appointed its first director and assigned the task of collection, collation and analysis of intelligence from beyond India's national frontiers. A 'Special Operations Branch' was included in its organisation for conducting secret and covert operations. RAW started with a staff of 250 persons and an annual budget of Rupees two crore. It expanded rapidly and by 1990 had a staff exceeding 8 thousand persons and a budget of Rupees 500 crore. Its present budget is around Rs. 1500 crore. It has its own aircraft and helicopters, the headquarters is located in large eleven storey building in Lodhi Estate, New Delhi. The present director is Mr. A. S. Siyali.

RAW has its own agenda particularly concerning neighbouring countries and has great influence in formulation of external policy of India. Since the last few years RAW has become so powerful and influential that it has emerged as a major player in the internal politics as well.

Besides the usual functions as an intelligence agency, RAW is relentlessly working for attainment of the following objectives. One, to extend Indian sovereignty over unprotected and vulnerable smaller neighbours. Two, to enhance Indian influence particularly in the Indian Ocean region militarily, politically and culturally, so as to project and develop India as a future super power. Three, to implement the Brahmanic dream of establishment of 'Akhand Bharat'.

The book explains how RAW's popularity, acceptability and influence increased manifold after its success in disintegration of Pakistan in 1971 and annexation of Sikkim in 1975. However, its image suffered some setbacks due to debacle in Sri Lanka during

1987 and its failure to have Pakistan declared a terrorist state. Nevertheless it retains its image as the country's premier intelligence agency wielding considerable influence in the formulation of foreign policy and internal security.

The book goes on to explain the cardinal principles of RAW inspired foreign policy. Which is to exert influence on other countries under the cover of friendship and co-operation: resort to threat and coercion to achieve desired objective when necessary; insist on policy of bilateralism etc. RAW attains these objectives abroad by conducting extensive espionage activities by the under cover diplomats and staff posted in Indian Missions abroad; recruiting leaders and other important persons; putting pressure or luring through incentives; brain washing intellectuals and spreading cultural influence; promoting internal clashes and developing separatist movements; influencing the government, members of parliament and government officials; forming pressure groups by spreading false, and distorted news; arranging assassination of important personalities etc.

The author explains the espionage set up of RAW in and around Bangladesh. Inside the country the prime centres are Indian High Commission at Dhaka and Deputy High Commissions at Chittagong and Rajshahi. Another RAW organisation operates around Bangladesh with its regional office at Malibagh, Calcutta. It has an annual budget of Rs. 20 crore and supervises the conduct of espionage and special operations inside Bangladesh. Its three regional offices are located at Darjeeling in the North Shillong, in the North East and Agartala in the East. Bangladesh is therefore very well covered from all sides.

RAW agents employed in Bangladesh can be divided into three broad categories. One, trained Indian nationals who come as diplomats, journalists, businessmen, students, cultural activists, literature etc. Two, Bangladesh nationals for a variety of motives including financial rewards, business interests, ideological considerations (some Hindus). Three, officials of multi-national organisations, NGOs, business houses etc.

The book vividly describes how RAW operated in the former East Pakistan and helped in the creation of Bangladesh. After that it has consistently tried to destabilise the new country politically and financially with the aim of merging it with India as planned by the Indian National Congress. This fact is well illustrated by a letter written by Mr. Nehru on 23 May 1947 to Mr Ashrafuddin, a congress leader from Comilla quoted by the author and attached as an appendix to the book.

'The Congress had stood for the union of India and still stands for it'. Wrote Mr Nehru, 'But we have previously stated that we are not going to compel any part against its will. If that unfortunately leads to a division then we accept it. But inevitably such a division must mean a division also of Bengal and Punjab. That is the only way to have a united India soon after. If we can have a united India straight away without such a division, that will of course be very welcome'.

This book is written by a highly educated man who was a student leader, a freedom fighter, a politician, and is now a journalist of great standing in his country. It is written with great perception and knowledge, quoting accurate facts and figures to illustrate his point of view. It should be read by our policy planners and all those concerned with the well-being and prosperity of South Asia. The book does show India's policy, present attitude and future plans concerning her small neighbours.

RAW USES FORD FOUNDATION IN ANTI-BANGLADESH CAMPAIGN

Jaideep Saikia, an outward analyst, but virtually an Indian intelligence operative, hailing from Assam, abruptly tunes to India's anti-Bangladesh campaign that the demography of Assam is being rapidly changed due to the alleged infiltration of the Bangladeshi Muslims into Northeast India, particularly in Assam in his recent book, "Terror sans frontiers: Islamist militancy in Northeast India".

Educated, better to say trained, in school at the Rashtriya Indian Military College in Dehra Dun, Saikia recently researched on so-called Islamic Militancy in North East India under the aegis of a Ford Foundation fellowship, which was awarded for the year 2003. The research was conducted at the Program in 'Arms Control, Disarmament, and International Security'(ACDIS) at the University of Illinois at Urbana-Champaign. It is an astonishing and utter folly how 'For Foundation' could sponsor and allow Saikia to use his fellowship on such an issue, which is not only controversial, but also baseless and false and a part of India's anti-Bangladesh media campaign.

Saikia's effort cannot be termed as research work, as this type of stories is written almost daily in India. India's electronic and print media, including websites, are poured with such fabricated anti-Bangladesh items. It is assumed that Indian intelligence outfit RAW(Research & Analysis Wing) managed and possibly financed 'Ford Foundation' to award fellowship to Saikia, which he used not only to defame Bangladesh, but also to prepare a ground for India to invade Bangladesh.

Without deep and careful study it can easily be questioned, how Saikia, being a researcher could write, like his all other fellow-Indians, an essay having minimal statistics and historical facts, which he on the other hand, distorted in every possible ways. He tuning to his mentors in New Delhi chorused that Bangladeshis deliberately infiltrate into Assam to change the demography of the state either to form a new Muslim state out of Assam or merge the Muslim majority areas of the state with Bangladesh. To justify his claim, Saikia says, "The Muslims now constitute more than 70 per cent of the population of Dhubri district of Assam. But Saikia did not mention from which source he collected this religion-based demographic information, as the Indian census of 2001, did not enumerate its citizens on the basis of religion.

Secondly, he should know that at least five districts of Assam adjoining Sylhet had Muslim majority in 1947, when the subcontinent was partitioned. These districts were Goalpara, Hilakandi, Cacher, Dhubri and Karimganj subdivision of Sylhet. For this reason, the Muslims constitute about 30 per cent of the population of Assam. So whatever might be the percentage of the Muslims in any district of Assam it cannot be termed as a threat to Assam or India.

Thirdly, Assam or any other state is not richer than Bangladesh, rather many states of India, not to speak of Assam lag far behind Bangladesh to a great extent. So why should the Bangladeshis leave for a poorer region to lead a poorest life. It is to be mentioned that Assam Gano Parishad, (AGP) is the prophet of anti-immigration crusade in Assam. But during its 2-term rules, AGP government under Prafulla Kumar Mahanto could identify few Muslims as illegal infiltrators in Assam. Even the current Congress Chief Minister Tarun Gagoi and Former Chief Minister late Hiteshar Saikia officially acknowledged that there is no illegal infiltration of the Muslims in Assam.

Meanwhile, the Ahoms, including the mainstream secessionist outfit ULFA (United Liberation Front of Assam), comprehended the design of RAW to divide the people of Assam into several antagonistic groups and crush them using one against the other mainly to frustrate the freedom struggle of Assam.

Realising the duplicity of Indian government, ULFA in July 1992 publicly declared the Bengali speaking migrants, which also include the Muslims, as friends. In a publication addressed to the 'East Bengal migrants' ULFA stated: "East Bengal migrants are considered Assamese. Without these exploited lot, ULFA cannot be successful. These are people who are educationally, economically backward. They cannot be our enemies.

These hardworking people are ULFA's protection shield. Their contribution to the national income is immense. They can produce essential things from a small piece of land, sell without any profit, work hard for the betterment of Assam, sacrificing them for the future of the state. They are our real well wishers, our friends, better than the Indians. ('The Revolution Comes Full Circle: Bibhu Prasad Routray.) In the same publication, ULFA went on to define the term 'Bidekhi' (foreigner).

"Those who do not regard this state as their own, accept it as their motherland, are not ready to sacrifice their lives for the sake of this country, are aliens, 'Bidekhis for us." Saikia should have read this statement of ULFA. He should also know that the Bengali Muslims accepted Assamese as their mother tongue and identify them as Ahoms not as Bengalis. The new generations of the Muslim Ahoms even do not know Bengali. They are not antagonistic to the interest of Assam. All these factors prompted ULFA not to brand the Muslims as foreigners.

Being failed to brand the Muslims as infiltrators or outsiders, very recently India floated another allegations that Bangladesh designs to secede the Muslim majority districts of Assam either to merge with her territory or create an independent Muslim state in Assam. Virtually, the campaign is made to create anti-Muslim sentiment among the Ahoms so that the unity among communities becomes far a cry.

Saikia and other Indians not only floated the allegation of infiltration of the Bangladeshi Muslims to Assam, but also allege that Bangladesh in one of the mentors of the decades old secessionist militancy in Northeast. According to the allegation, which Saikia also did not forget to forefront in his book, Bangladesh provides shelter, training and even arms to different militant groups of the region, particularly ULFA, ATTF (All Tripura Tiger Forces), etc.

But being an Ahom and above all a researcher, Saikia should know that secessionist insurgency in Northeast when India got its independence from Britain in 1947, well before the birth of Bangladesh. People of this region do not feel them as Indians. They are fighting to end what they call, "Indian occupation." Previously India blamed China, Burma (Now Myanmar), Pakistan and even America. But they shortened their list over the years and ascribe the allegation on Bangladesh and Pakistan. Some of the Indians now consider Bangladesh more dangerous for northeast than Pakistan. This allegation against Bangladesh was brought to the forefront, because it will be easier to squeeze weaker Bangladesh than any other country that India blames

But India could never prove any of her allegations against Bangladesh. India officially challenged that there are 195 camps or training centres of the Northeast insurgents in Bangladesh and supplied a list mentioning their whereabouts. According to the list, training centres and camps are situated in hospitals, police stations, residential colonies, government offices, playgrounds, etc. Bangladesh repeatedly requested India to come and show on-spot the existence of these camps and centres. But India never accepted the offer, as Indian policymakers know that there is not even single such centre or camp of the northeast militants, not to speak of 195.

Still the propagandists in New Delhi deliberately continue their fabricated allegations against Bangladesh, whose brief ulterior reasons I have mentioned earlier. I really feel pity for Saikia as well as Ford Foundation for being used as the tools of RAW. How Ford Foundation could accept such a baseless research work which goes against a country, which is a main target of Indian expansionist design. I would request Ford Foundation to send a 'fact finding mission' to Assam and Bangladesh as well to inspect the ground realities. Such mission will surely find that all the allegations that Saikia mentioned against Bangladesh in his so-called research work are the products of exaggeration and misinformation. Ford Foundation, to uphold its neutrality and worldwide reputation and acceptability, should consider my suggestion and act accordingly.

August 17 Blasts: Is there external linkage?

The controversy over Tarique Zia's seemingly misquoted comment in the BBC interview that al-Qaeda 'may' have been involved in the August 17 serial blasts notwithstanding, despondency is bound to set in as the investigators have not yet unearthed any significant leads to the attacks' masterminds (and their political goals) despite over 300 arrestees' testimony having been recorded and a slew of clues found. Such uncertainty does give rise to an obvious concern: Is there an external link to the blasts and, if so, who could have pulled the strings from behind the nation's borders, and why?

A study of the post-blast behaviour of the Indian media and the intelligence apparatuses can go a long way toward understanding why terrorist incidents in Bangladesh seem to matter so much to our neighbour. Since the attacks, the Indian media has launched a virtual crusade against Bangladesh, spearheaded by the Telegraph that wrote, 'Delhi should urge major donors to impose economic sanctions on Bangladesh.' The paper also reported that Indian security agencies had advised the central government to 'force Khaleda Zia to clamp down on Islamic fundamentalist outfits'. The government of Bangladesh did respond earnestly to such pressures and diatribes and conducted a virtual witch-hunt in the preceding weeks against Islamists of suspicious hues, although the end result of the ongoing manhunt seems destined to be as much a failure as the previous ones.

Meanwhile, a just concluded study of Bangladesh's post-blast security situation by major Indian intelligence outfits pointed the finger of suspicion for the August 17 blasts at familiar groups like the Jagrata Muslim Janata and the Jamaatul Mujahideen, which are, says the study, 'banned, and are known to have fanned anti-Indian sentiments'. Coincidentally, the police in Dhaka say the same thing but cannot trace the attacks' elusive masterminds.

The masterminds of a series of such attacks over the years not having been traced, one cannot resist the temptation of being suspicious about the latest attacks' genesis and the ultimate motivation of the masterminds.

The Indian intelligence bodies' study, however, has made some interesting observations. 'There were 370 explosions in 63 of Bangladesh's 64 districts. The kind of explosives used and the impact of each blast were similar to that on August 13 at a Muslim shrine in Akhaura in which one person was killed and 30 others were injured…There are insinuations that an earlier blast in August 2004 was suspected to have been inspired by India's Research and Analysis Wing (RAW). It is possible that efforts will be on to malign India again by pointing fingers at Delhi,' the study opined.

In conclusion, the study noted that 'the outfits were emboldened because of the lack of tangible action by the Khaleda government… The blasts are intended to be a message to Dhaka and to discourage the government from succumbing to international pressure to clamp down on the outfits.'

Reportedly a separate detailed study, circulated among the top echelons of the Indian security establishment, says, 'Delhi should actively consider economic measures against Bangladesh.'

The tirades of the Indian media and comments of the intelligence agencies aside, everyone knows that the Jamaatul Mujahideen had left leaflets on the sites of bombing and many of the arrestees have reportedly confessed to having carried out the attacks at its behest. Yet BNP Deputy Minister Ruhul Quddus Talukder (also an MP) had a different view of the events. He had said earlier, 'I don't think they (the JM) have such a strong network. Awami League must have done this, using fake leaflets, to blacken Bangladesh's image internationally.' Does the Minister know something that others don't?

A somewhat similar claim came from Mufti Fazlul Haq Amini, chairman of the Amini faction of the Islami Oikya Jote and a constituent member of the ruling four-party alliance. He said on August 19, 'Swearing upon Allah, I say the 14-party alliance of Awami League and left parties launched the bomb attacks in a planned way to uproot the Islamic forces, but Islamic forces can never be eliminated.'

To confound confusion further, both India and Israel were whisked into the scene by the Jamaat-e-Islami's Amir and Industries Minister Matiur Rahman Nizami. He blamed India's external intelligence agency, Research and Analysis Wing (RAW), and Israel's Mossad for 'playing an important role' in the August 17 attacks. He added, 'They are the patrons of the serial blasts as they don't want good relations between Bangladesh and China. That's why the incident occurred when Prime Minister Khaleda Zia was on a visit to Beijing.'

Juxtapose the above with the embedded Indian concerns over Bangladesh's political developments over the years. The copy of a 2004 RAW report obtained by this author reads, 'Pakistani intelligence officers in Dhaka are becoming increasingly active in espionage against India. In 2002, three modules (sic) being run by them from Dhaka, and using some Bangladeshi operatives, were busted. A large number of secret documents and photographs of sensitive defence locations were recovered from one Ziauddin Ahmed Biswas (resident of Murshidabad in West Bengal), arrested on November 17, 2002. Later, the arrest (December 2002 in Lucknow, UP) of Bangladeshi national, Mohammad Mamunur Rasheed, led to the recovery of fake travel documents and also incriminating documents indicating a plan to recruit Indian Muslim youths for training in Bangladesh and Pakistan for subversive activities within India.'

Another RAW report of 2004 implicates the Dhaka regime more directly. It says, 'It is hardly any secret that the Inter-Services

Intelligence (ISI) of Pakistan has close links with Bangladesh's Directorate-General of Forces Intelligence (DGFI) and operates openly and freely in that country. It (ISI) not only helps coordinate the activities of al-Qaeda and fundamentalist Islamic militant groups through the DGFI, but backs a Bangladeshi Taliban group named HUJI that runs six training camps for ULFA terrorists in the Chittagong Hill Tracts.' A West Bengal intelligence outfit goes a step further: 'While ULFA training camps have been organised by the sector headquarters of the Bangladesh Rifles (BDR), training camps of the CNLF have been organised partly by 103 and 105 infantry brigades of the Bangladesh Army at Khagrachhari and Rangamati,' the report claims. From hindsight, the pattern of such accusations seems a corollary to many such reports circulated in the past. For instance, prominent security experts of India have been crying wolf since the late 1990s (long before the incidents of 9/11 that acted as a harbinger to the global hunts for Islamist terrorists) that activities in Bangladesh posed a serious danger to India's security and national interests.

Particularly, Assam Governor Lt Gen (retd) SK Sinha wrote in his report to the central government in March 1998, 'The long cherished design of Greater Bangladesh, making inroads into the strategic land link of Assam with the rest of India, can lead to severing the entire land mass of the North East from the rest of the country.' In another report submitted to the President of India in November 1998, Sinha wrote, 'Continued silent demographic invasion of the North East poses a great threat both to the identity of the Assamese people and to our national security.' Influenced by such reports, India decided to fence the entire Indo-Bangladesh border at a cost of over $500 million and nearly 70% of border fencing was completed by mid-2005. The Indian Border Security Force also killed more than 500 innocent Bangladeshis over the years since General Sinha filed his first report.

India now claims that since 1990, Assam has seen the birth of 9 Muslim militant outfits owing allegiance to Harkat ul Mujahideen and Lashkar-e-Toiba, the groups that run ferocious operations against Indian forces in Indian-occupied Kashmir. Indian intelligence outfits believe the groups have their rear bases inside

Bangladesh. Is India looking for a pretext to launch pre-emptive military assaults on Bangladesh at some point in the future, based on such reports? Policy-makers in Dhaka must mull over this prospect seriously.

Coming to the August 17 blasts in particular, one wonders why the Islamists, whose 'profound' aim is to create a 'Greater Bangladesh' by creating demographic imbalance in the neighbouring Indian states of Assam and Tripura in particular (according to Indian reports), should resort to blasting of 'innocuous' bombs inside Bangladesh and leave behind signatures for identification? How is the mission of creating a greater Bangladesh served by such blasts? Isn't it more plausible that, in the absence of any verifiable and authentic conclusion, the blasts have occurred to prove to the world that Bangladesh is infested with Islamist Jihadis determined to take on India by using Bangladesh as a launching pad? At the least, such a hypothesis does mesh well with the embedded Indian perceptions of Bangladesh, as has been learnt from the intelligence reports quoted above.

It is under such contexts that one must compare the Indian mindset with the comments made by some Bangladeshi politicians after the August 17 blasts (quoted above), and try to guess the 'untold' reasons behind the authorities' inability to reach any conclusion with respect to the attacks' masterminds.

5

Intelligence Agencies in India

Today's intelligence agencies operate in highly complex environments. Cold War definitions and understanding of threats have long become redundant. Threats are multiple, layered, networked, diffused and transcend social and spatial boundaries. Traditional distinctions between domestic and global threats have greatly blurred, especially in the case of terrorism. In this sphere, most pressing challenges for any intelligence agency is to counter, contain and interdict stateless violent actors and state-supported violent actors. This has introduced some fundamental changes in the manner in which intelligence is collected, analysed and disseminated, how intelligence agencies operate, collaborate and at the same time protect their strategic and tactical instruments and objectives from any influence and compromise, accidental or otherwise.

Although intelligence agencies in India have been slow to transform, primarily because of bureaucratic inertia, poor leadership and lack of political direction, significant changes have been witnessed in the intelligence community, particularly after the Kargil conflict of 1999, and more recently after the Mumbai attacks of November 2008. In comparison to the past, the intelligence agencies are today much more tech savvy, better staffed, well coordinated and have benefited from cooperation with other international agencies in training, orientation and sharing of intelligence.

But the agencies, particularly the Intelligence Bureau and Research and Analysis Wing, have shied away from one of the best practices followed in most of the democratic world's intelligence community. That is to institutionalise a robust interface between the agencies and the Indian citizen.

Scope and definitions

This chapter argues that intelligence agencies need to interact with the citizens at multiple levels in the overall national interest and there are compelling reasons for doing so, in terms of seeking a public mandate for the activities of intelligence agencies in a democratic system and for introducing transparency in the functioning of the agencies for better accountability, and performance.

At this stage, it is important to define terms like 'intelligence agencies' and 'public' with more clarity.

'Intelligence agencies' here means the Intelligence Bureau (IB) reporting to the Ministry of Home Affairs and Research and Analysis Wing (RAW) working under the overarching jurisdiction of Cabinet Secretariat. The heads of both the agencies, however, have the right to report directly to the Prime Minister. Both can also be asked to report to the National Security Advisor. The term 'public' is more broader in its meaning and includes the people's representatives, Parliament, academics, business and industry leaders, journalists and other citizens.

A brief look at the overall security/intelligence architecture in India as it exists today could also be relevant to our enquiry. The intelligence architecture rests on four organisational pillars—political, administrative, intelligence and enforcement. The Cabinet Committee on Security headed by the Prime Minister with External Affairs Minister, Home Minister, Finance Minister and Defence Minister as members constitutes the political arch of the system. The National Security Adviser acts as the ex-officio member of the committee. The administrative part comprises the Ministry of Home Affairs, the Prime Minister's Office and the Cabinet Secretariat. The intelligence wing constitutes of Intelligence Bureau (reporting to the Home Ministry), Research and Analysis Wing (reporting to the Cabinet Secretariat which means to the Prime Minister), Joint Intelligence Committee, National Technical

Research Organisation and Aviation Research Centre (reporting to the National Security Advisor) and the National Security Council Secretariat (under the National Security Advisor). The military has its own intelligence units and an overarching body called the Defence Intelligence Agency. Besides, there are several financial intelligence units working under the Finance Ministry. The enforcement is the responsibility of a host of police and para-military organisations.

This paper's scope, however, is confined to the intelligence agencies and the manner in which they interact with the public, and explore possible ways to enhance this relationship in the overall interest of national security.

Twitter world

In a highly globalised world where social media networks are the new medium of dramatic changes, the intelligence agencies in India continue to function in isolated splendour, insulated in the secure confines of past traditions and mindset. Two reasons agitate against such behaviour. The need to know has become a sheet anchor of transparency in governance and this bourgeoning democratic urge can only be ignored at some cost to the democratic structure of nation's polity which the intelligence agencies are committed and tasked to protect. There is hardly any incentive for a professional organisation to bury its head in the sand when the world around is shifting rapidly.

Second, and this is in a way more important for an intelligence agency, public awareness and commitment could act as 'force multipliers' in many ingenuous ways than earlier imagined, thanks to the technological and information waves sweeping the world. State of isolation drives the intelligence agencies away from being an important and integral part of India's comprehensive national power. Instead, they become more secretive which, as is often the case world wide, spawns various levels of malpractices and failures. These 'failures' then induce the intelligence officials to seek political patronage and refuge, further undermining the professionalism within the forces. A few illustrative case studies reinforce the above argument. The 1999 Kargil conflict brought serious charges of failure on the part of intelligence agencies although it was known that there were adequate intelligence inputs about troop movements, increased infiltration and other associated events. The failure on the part of the intelligence agencies was more in connecting the dots they had collected over a span of time. But this failure was accentuated by the army's own set of failures in correctly assessing its own inputs as well as the ones shared by both IB and RAW. These mistakes were further compounded by unpardonable political hubris. The intelligence agencies found no viable platform to contradict or clarify the allegations against them.

A similar situation arose during the Mumbai 2008 attacks. The failure of the intelligence agencies to forewarn, and pre-empt, the attacks raised public hackles once more. Once again, it was a systemic failure, including that of intelligence agencies. The failings of the law enforcement agencies, the federal command and control structure and the political leadership were too many. While the bureaucrats and the ministers used the media to express their views, largely to defend themselves, the intelligence agencies had no chance to explain their stand.

A charitable view, which has a large following within the intelligence community, could be that such a position also helped the agencies in deflecting criticism. The flipside of such an argument is that the cover of anonymity which the agencies enjoy in India not only makes them immune to public scrutiny but also to introspection and to effect necessary reforms to strengthen their professional duty towards the nation.

Missing interface

Democratic governments disseminate policy decisions, guidelines and directives through official spokespersons. This 'public affairs department', often staffed, in India, by members of the Indian Information Service (IIS), varies in size depending on the department. For instance, the Ministry of Defence has a Directorate of Information and Publicity headed by a senior IIS officer who is authorised to brief the media, issue press statements and organise press conferences on behalf of the Defence Minister, his deputy Ministers, Defence Secretary and various heads of defence-related departments. He also oversees the functioning of the information departments of Army, Navy, Air Force and Defence Research and Development Organisation (DRDO). The military units are staffed by respective members of the Services reporting to the civilian head of the Directorate. The Army also runs a Army Liaison Cell which interacts with the media but not on a regular basis.

In the case of intelligence agencies in India, there is no institutional mechanism to inform the Indian people about their functioning and utility. In theory, IB should be represented by the official spokesperson of the Home Ministry and RAW by the Cabinet Secretariat or the Prime Minister's Office. In practice, neither of these have ever dealt with intelligence matters. The IB's parent organisation, the Ministry of Home Affairs, has an official spokesperson but most of the briefings to the media are done either by the Home Secretary or the Home Minister. The parliamentarians are briefed by senior officers of the ministry, including the Secretary. On rare occasions, senior IB officials have directly briefed the parliamentarians. The Prime Minister's Office is represented by the media adviser but rarely has this office held any briefings on intelligence issues.

As an alternative, the National Security Council Secretariat could have taken the responsibility of communicating to the people about domestic as well as global threats impinging on India's overall security. But in the absence of an official spokesperson, the Council's assessments are made public through talks/speeches of the National Security Advisor.

The intelligence agencies, however, have instituted an ad hoc mechanism to converse with the media. In RAW, a senior official 'feeds' the media while IB has a similar group of officers who interact with the journalists. Occasionally, certain unclassified materials are 'leaked' to the media. A recent case was that of a list of Lashkar-e-Tayyeba leaders who were named in a US court for the Mumbai attacks. Several years ago, RAW had prepared an unclassified compendium of terrorists operating from Pakistan which was then circulated among select few journalists, academics and other opinion makers. In essence, the relationship between the agencies and the wider public is at best opaque and intermittent in nature.

Similarly, the debate on intelligence matters in the Parliament or in the various standing committees has rarely been informed by an understanding of the complexities of an intelligence operation. Most parliamentarians learn about intelligence matters from newspapers and magazines. Only a very few correspondents and editors in India have studied intelligence matters or agencies to write with any authority on the subject. As a result, barring occasional sensational leaks and conspiracy theories, there is little information about the intelligence agencies in the mainstream newspapers and magazines. This cycle of ignorance gives rise to indifference, doubt and suspicion in the minds of the citizens about the intelligence agencies.

Need to know

The 'need to know' principle is stretched to a ludicrous extent in the Indian context. The intelligence agencies have never published any threat assessments and threat perceptions. The only public media where threat perceptions are published are the Annual Reports of the Ministry of Home Affairs and the Ministry of Defence. In comparison, the US and several European countries regularly publish intelligence estimates, threat assessments besides declassifying a vast amount of confidential notings and reports every year which help in building a transparent, easily accessible public database on intelligence agencies and their outputs. In India, the official threat perceptions can only be gleaned from the statements, interviews and articles written by senior ministers or bureaucrats.

In fact, the intelligence community or its members do not publish anything. They also discourage, barring a few exceptions, publishing of memoirs or accounts by retired officers. Except for accounts of their careers in intelligence services written by Mr B. Raman (former Additional Secretary, RAW) and Mr Moloy Krishna Dhar (former Joint Director, IB), and an expose of sorts by former RAW officer, Major General V.K. Singh, there is a dearth of published material from the intelligence officials. Some of them do take to writing opinion pieces after their retirement but those provide hardly any clue about the nature and functioning of intelligence agencies. Reports of committees that studied reforms in the intelligence agencies remain either classified or are available in parts, that too heavily redacted. The Kargil Review Committee report was, perhaps, the last such publication.

The intelligence communities in India do not have an internet presence either. Unlike in the case of CIA or MI5, the Indian intelligence agencies do not host any website of their own. Neither does the website of the Home Ministry nor that of the Prime Minister's Office offer any clue about the intelligence agencies. The National Security Council remains without a web presence while the Cabinet Secretariat website is silent on the agencies. Thus, the Indian agencies are virtually non-existent in the networked world.

CIA's Outreach Programme

On the other hand, the online as well as offline presence and visibility of the intelligence community in the US is admirable. This is primarily due to the heavily invested and empowered offices of 'public affairs'.

A brief overview of the 'public face' of the Central Intelligence Agency, perhaps one of the largest, technically empowered intelligence agencies with an enormous budget, can provide an indication of the importance given to public interaction. The CIA's public interaction is conducted by the Office of Public Affairs (OPA) and its Director acts as the CIA spokesperson. He reports to the CIA Director. His responsibility includes "communicating with the media, the general public, and CIA's workforce". The OPA in fact acts as the eyes and ears for the agency and keeps the top leadership "informed of prospective and breaking news stories about CIA, including stories that could put Agency officers or operations at risk". The office develops media strategies, facilitates the Director's press briefings, interviews and public appearances. The CIA's public website remains a storehouse of declassified material, easily searchable, and a host of publications, including the widely popular and widely desseminated online CIA World Fact Book.

The CIA's outreach programme has been in existence since 1947 and has been subjected to continuous reviews by in-house assessments as well as Presidential Task Forces. The Task Force on Greater CIA Openness set up in 1991 made significant recommendations on making CIA transparent and responsive to the public quest for information. One of the questions it set about answering was—How can we do a better job of informing the general public and key constituencies about the need for a strong intelligence effort and about the missions and accomplishments of the intelligence community in a changing world?

One of the conclusions the Task Force reached was that there was a consensus, among those who were consulted, on the "need to make the institution and the process more visible and understandable rather than strive for openness on specific substantive issues. To do this, we need to develop strategic vision of what we want to be open about, why we want to be open and to whom we want to be more open".

BRITISH INTELLIGENCE AGENCIES

The British example could also help in understanding both the need and the dilemma of opening intelligence agencies to public scrutiny and dissemination. Two key intelligence agencies of the British government are the Secret Intelligence Service (SIS), more widely known as MI6 and the Security Service, better known as MI5. The first one is tasked to provide "Her Majesty's government with a global covert capability to promote and defend the national security and economic well-being of the United Kingdom'. The latter is "responsible for countering covertly organised threats to national security". MI6 does not maintain a press office but has a designated press officer at the Foreign and Commonwealth Office to respond to queries, and a website. The chief of MI6 produces a classified report for the Prime Minister as well as the Foreign Secretary. Parts of this report are available in the public domain through reports published by the Intelligence and Security Committee. The domestic intelligence agency, MI5, has no press office and handles media queries through the Home Office. But it maintains a website to "improve public understanding of our role". The website, it is stated, "contributes to the government's counter-terrorist strategy by providing a detailed description of the threats to national security and the current national threat level". The site also helps public to send "us information they think could help with our work".

Despite these measures at 'openness', the British agencies have been grappling with the problem of 'media leaks'. The 2008 Annual Report of the Intelligence and Security Committee cited cases of misreporting and media leaks which came 'close to compromising counter-terrorism operations' and even, in some cases, damaging the long-term capability of the agencies. The committee noted that the role of media in counter-terrorism was "more important than ever'. It said "the current system for handling national security information through DA notices (an unofficial media advisory to avoid publishing stories that undermine intelligence/security operations) and the agencies' relationship with the media more generally is not working as effectively as it might and this is putting lives at risk. We recommend that the government engage with the media to develop a new, effective system, with a view to protecting intelligence work, operations, sources and criminal prosecutions while ensuring that the media continue to report on important matters of public interest".

Significant Payoffs

The extent of debate in the intelligence community about the need to manage media projections and public interaction. There is a general consensus, with some caveats, about the need to convey certain messages to the people the intelligence agencies are tasked to serve and this can only be achieved by establishing a broad framework of interaction either directly through a press office or indirectly through websites, external media office, oversight committee reports and publications.

There are obvious outcomes of better public- intelligence agency interaction. These include a greater public awareness about threats which can help the agencies as well as the law enforcement agencies in tracking extremist/terrorist groups, hideouts and other suspicious activities. An active and aware resident welfare association or a community group, if suitably aware, can be of immense help in gathering ground level intelligence. Sharing information about threats to infrastructure, market places and mass transit systems like Metro with the people through media, background briefing to local elected representatives, local police and civic officials could make terrorist activities difficult to carry out.

Another positive outcome of a better interaction would be in the quality of candidates attracted to the services. Recruitment to intelligence agencies, like their other activities, remains no less secretive. Allegations of nepotism are rampant in the agencies. The overall quality of personnel remains questionable. The recruitment of top notch students from social science, engineering, computer professionals, economists and media into intelligence agencies remain almost negligible.

Moreover, public perception of threats can influence public policies. India's enormous investment in homeland security after the Mumbai attacks has been provoked in considerable measure by the enormous public outcry and anxiety. Such incidents also provoke unbridled, quite often misplaced, criticism of the intelligence agencies, clubbed in popular parlance as 'intelligence failure'. An attack of the magnitude of the Mumbai attacks is a multiple, systematic failure and to lay the blame at the doorsteps of the intelligence agencies is unreasonable and in some ways counter-productive. But these events create enormous public upheavals, resulting in uncalled for criticism and scrutiny of the intelligence agencies. Though without a sound reasoning, such criticisms morph often into beliefs and undermine the morale and efficiency of the intelligence staff.

The intelligence agencies, by their very nature of functioning, shy away from public dealings and scrutiny. This excessive diffidence is strengthened by media leaks, sensational stories and wrong, misleading reports involving the agencies. The government's decision in 1999 to make public the transcript of conversation between General Perverz Musharraf and his aide, Lt.

General Mohammad Aziz Khan, at the height of the Kargil conflict, had not gone down well with the intelligence community. It is said that the conversation between Musharraf in China and Aziz Khan in Islamabad was tapped by RAW by intercepting a particular satellite link. By making the conversation public, some officials believe, a key source of SIGINT (signal intelligence) was compromised, affecting future intelligence gathering operations. An incident of this nature, in the Indian context, has been more of an exception and hence chances of its repetition are rare, if any. The issue of misreporting and wrong projections by the media and commentators are a problem which occurs often and can be corrected by effective communication strategies.

Modalities of interaction

There are already some established modalities of interaction between intelligence agencies and the public. These include media briefings, press statements, publication of annual reports and other assessments, if any, and speeches of ministerial heads. Deep background briefings for the media have been a key instrument to inform journalists and their writings. Websites offer another public platform for the agencies to interact with a wider section of the public. In the Indian context, none of these interface mediums are in practice. The agencies can, however, draw lessons from the experience of the armed forces in this regard. The Indian armed forces—Army, Navy and Air Force—have their own official spokesperson and website to project the views of the military leadership and promote its activities to a wider audience. Besides, the armed forces are also served by a civilian directorate of information and publicity located within the Ministry of Defence whose head reports directly to the Defence Secretary and Defence Minister. The Defence Ministry publishes Annual Reports and other publications which provide an insight into the functioning of various wings of the military establishment. The Standing Committee on Defence publishes regular reports on various aspects of the military, in addition to the regular audit reports prepared by the Comptroller and Auditor General of India. In case of the intelligence agencies, there are no such reports available in the public domain. Although there have been financial audits of RAW and IB, the reports have remained classified. Since there are no oversight parliamentary committees on intelligence, the functioning of the intelligence agencies remains opaque and dense.

Recommendations

The government must set up a task force with one representative each from the intelligence community, media, political party, academia/think tank and the National Security Council Secretariat to suggest a framework for institutionalising a more transparent public-intelligence interface. The task force must seek answers to the following questions: Is there a need for an official spokesperson or media office for the intelligence agencies? If so, where should it be located? If not, can the National Security Council Secretariat house a public affairs office to disseminate information on matters of intelligence, provide background briefings to the media and respond to media queries? Should unclassified threat assessment be published separately? Or should it be part of the Home Ministry/Defence Ministry Annual Reports? Is there now a need to communicate with the internal staff of intelligence outfits and what is the most effective manner to do this? Should transparency in dealing with threat to the nation not be a cornerstone of a strong and confident democracy?

INTELLIGENCE REFORMS

The need for reforms in the Indian intelligence set-up does not spring from any desire to ape the West, but from the fact that notwithstanding numerous failures so far, there does not seem to have been any broad-based exercise to reform the country's intelligence apparatus and make it more pro-active and in harmony with the pursuit of nation's internal and external policies. Whatever piecemeal restructuring that has been tinkered with from time to time, has mostly been crisis- driven and not a comprehensive need-based attempt to address the basic and structural flaws in the Indian intelligence set-up that appear to be a legacy of its origin as a branch of policing of the society.

The Indian intelligence system did not evolve out of any detailed and well thought out administrative policy, but emerged as an extension of the Indian police system due to a need driven colonial decision making process, designed to meet the requirements of maintaining law and order and internal security. Its primary duty was to keep a check on the nascent ideas of modern Indian nationalism and counter them through police methods that were considered most effective and adequate.

According to Shri M. Hamid Ansari, the Vice President of India, "While intelligence information is at times incomplete, good intelligence often has made the difference between victory and defeat, life and death. By the same token, faulty intelligence leads to failures of varying degrees". Despite some success stories, the Indian intelligence agencies appear to have failed on numerous occasions to perform by producing meaningful actionable and timely internal and external intelligence to safeguard national security and interests.

The Bangladesh operations, the advance information given to Indian Air Force regarding Pakistani plans to launch a pre-emptive air-strike on Indian bases on or before December 3, 1971, the unearthing of plans by many modules of terror groups to hit at Indian interests both inside and outside the country, are some of the major successes of the Indian intelligence community. Yet, these have mostly been like flashes in the pan and not reflective of the general levels of performances. Even in those cases, where Indian intelligence apparatus has succeeded in busting terror and espionage rings, it has not been able to provide accurate information which could be developed into concrete evidence able to stand judicial scrutiny (as was the case with MI-5's unearthing of the plot to blow-up trans-Atlantic flights over American cities).

Some of the major failures of Indian intelligence over the years have been:

1. Almost total ignorance about Chinese intentions and capabilities in Tibet and their designs against India in the 50s. (A case in point is the non-use of Indian Air Force to bomb Chinese positions on Indian side of the border during the 1962 Sino-Indian War, because of the fear that it could lead to Chinese retaliatory air raids against Indian cities in North India. This was a major intelligence failure since the PLA Air Force did not have large bases or aircraft in

Tibet at that time, having the capability to launch raids across the Himalayas, bomb Indian cities and return to their bases);

2. Failure to assess the extent of popular alienation Centre's policies were creating in Kashmir through repeated changes in state governments since the early 1950s and, more specifically, during 1986-87;
3. Failure to predict Pakistani plans to infiltrate military personnel disguised as civilians into J&K in 1965;
4. Total ignorance about Pak plans to attack Chhamb in the 1965 war and the existence of an additional armoured division in the Pakistani Army;
5. The extent of the growing unpopularity of Sheikh Mujib government in the Bangladeshi armed forces and the plans for his ouster;
6. Pakistan's decision to initiate its nuclear programme in 1972 (The Multan conclave of Z.A. Bhutto, Pakistani physicists and nuclear scientists);
7. The extent of popular support enjoyed by movements like those for a separate Telangana state, or the growth of left-wing extremism in the post-Naxalbari phase;
8. The Samba spy scandal;
9. Pakistani plans to subvert Kashmiri Muslims in the name of religion in the 70s.Pro-Deobandi/W ahabi proselytising activities in the Valley in the mid-80s and recruitment of Kashmiri youth for arms training in Pakistan and their infiltration back into the Valley with arms to launch the 'Kashmiri Jehad' in late 80s;
10. The developing situation in the North- east from the early 80s; Links between Phizo and the ISI;
11. The growth of Sikh-separatist sentiments in the early 80s and, more specifically, the volume of the arsenal stockpiled inside the Golden Temple complex before 'Operation Blue Star';
12. LTTE's reaction to Indian peace making efforts in Sri Lanka and the extent to which its armed capabilities had grown to target Indian interests;

13. Kargil incursions; involvement of Pakistan Army regulars from Northern Light Infantry (NLI) in infiltrating deep into Indian territory, and construction of fortified 'sangars';
14. Various terrorist attacks in India either directly by Pakistani players, or their Indian proxies, specially the Mumbai attack of 26/11 and those in New Delhi before that.

There are instances of intelligence failures of such magnitude in the developed world also. However, they have all been thoroughly enquired into, responsibilities fixed and remedial measures taken. However, this has not happened in India and the public impression persists that on most occasions the dirt was merely pushed under the carpet. After the Sino-Indian war of 1962, a review of performance of the Intelligence Bureau (IB) led to the creation of the Directorate General of Security (DGS). The Indo-Pak War of 1965 and the Mizo revolt in 1966, lead to the responsibility for external intelligence collection being taken away from IB and assigned to the newly created Research & Analysis Wing (R&AW). After the Kargil operations in 1999, an enquiry by Kargil Review Committee lead to the setting up of the G.C. Saxena Special Task Force which recommended the creation of the Defence Intelligence Agency (DIA) as the nodal point for processing all military related intelligence. The National Technical Research Organisation (NTRO) was also created to collect technical and communications intelligence. In June 2009, the Pradhan-Haldar-Narsimhan task force in the National Security Council (NSC) made suggestions for improving intelligence organisations in India. The Federation of Indian Chamber of Commerce and Industry (FICCI) also came out with a task force report on National Security and Terrorism, which made several suggestions for strengthening intelligence collection efforts and national security management.

However, most of these reports were crisis- driven reviews, retrospective in nature, rather than need-based broad perspective plans to reform and revamp. They looked into what went wrong, examined particular lapses and suggested ways and means to prevent their recurrence. More often than not their suggestions were for setting up new agencies to plug the gaps, such as the one being presently discussed for setting up a National Counter-

Terrorism Centre (NCTC); a new Maritime Intelligence and Coastal Security Centre; or a new Centre for Nuclear and Missile Intelligence, without clearly spelling out the ways and means to prevent overlaps or turf-wars. This persistent failure is, perhaps, significantly responsible for the tussle between the newly created NTRO and the Aviation Research Centre (ARC) over the use and bifurcation of expensive valuable assets and facilities. There is clearly a growing perception in the media and amongst an ever-widening spectrum of intellectuals and academics engaged in the study of national security related issues that a rigid and stodgy bureaucracy may have stood in the way of developing or enhancing the desired core competence in the field of intelligence operations and analysis. According to a leading intelligence expert, Indian intelligence set-up is like an aircraft that can fly at a maximum height of only 10,000 feet. If the nation has a need for an aircraft to go up to the height of 15,000 feet, no amount of overhaul and up-gradation of the old aircraft's systems would make it attain that height. There would be no alternative then but to develop an entirely new technology and have an aircraft that can go up to 15,000 feet and beyond.

However, apart from the natural inclination not to risk or gamble, three factors seem to have conspired against the adoption of an imaginative and unconventional approach to equip and improve capacities of the intelligence agencies to meet new threats, commensurate with India's emerging responsibilities as a global power:

i) Conflicting motivations of those considering reforms;

ii) Environmental challenges at initiation of reform; and

iii) Failure of leadership.

Understanding these factors and seeing where gains have been made despite handicaps may lead to the realisation that real change may still occur, but only if difficult choices are made while opportunities exist. Any quest for issues uncomfortable to the intelligence establishment in the country, such as, whether intelligence should remain the preserve of the police; whether recruitment and lateral intake policies should not be adopted to take advantage of the available immense cross–cultural talent in the form of political analysts, legal experts, linguists, financial wizards, social scientists, journalists and domain specialists, who could be roped in to provide valued inputs based on the acuity, ingenuity and contextualised insights of their own experience and wisdom.

A broad consensus that is apparently shared by a wide section of Indian intelligence professionals presently, underlines the need for the following:

i) A focus on "improvement of capabilities, or removal of deficiencies, within the agencies themselves, whether in terms of manpower or quality of assessments";

ii) "Co-ordination amongst various agencies- there must be someone who knows everything that is happening and who can go and report to the Prime Minister frequently", i.e. a "Minister for National Intelligence";

iii) "Thirdly, how do you manage meaningful reform in the intelligence machinery would necessarily bring up Parliamentary oversight?"

Shri M. Hamid Ansari has expressed concern about the nature and extent of supervision over intelligence agencies by the political executive and the possibility and scope of misuse of these agencies by that authority. He has also highlighted the issue of accountability and oversight as these measures were being undertaken in various modern intelligence organisations. Clearly, the time has come for the political executive to confront these issues head on and not skirt around them.

INDIAN INTELLIGENCE AND THE KARGIL CRISIS

During the summer of 1999, India and Pakistan fought a 10-week limited war in kargil, a remote area of kashmir. Fighting broke out in May, when Indian troops discovered that a number of armed men had crossed the Line of Control (LoC) and entrenched themselves on the Indian side. Over the following weeks, the Indian army learned that these gunmen were not Islamist guerillas, as it had first assumed, but Pakistani soldiers in mufti. A security crisis erupted, with allegations of 'failure' being thrown at the Indian intelligence agencies.

Initially, critics of the intelligence community believed that it had failed to predict the Pakistani offensive. They assumed that forewarning would have prompted new Delhi to discontinue its then-ongoing rapprochement with Islamabad. However, an official commission of enquiry (set up after the crisis) cast doubt on this view. The enquiry established that the intelligence community had been sceptical about the rapprochement policy and had conveyed as much to the political leadership. The surprise at kargil did not stem from what happened, but rather, an incorrect forecast of *how* it would happen.

Intelligence analysts had predicted a surge in cross-border infiltration by Pakistani and afghan 'mercenaries' (a term which has since been replaced with '*jihadists*'). Their assessment implied that, despite the peace talks, bilateral tensions would escalate. It did not anticipate that the Pakistani military itself would attack across the LoC. such a move was judged to be 'irrational' and was discounted as a possibility by both intelligence analysts and Indian military commanders. Instead, threats from Pakistan were conceptualised in familiar terms, and perceived to unfold in linear progression.

Following the publication of the official enquiry's findings, criticism of the intelligence agencies was recast from strategic to tactical levels. The agencies were now blamed for not correctly identifying the kargil intruders as being Pakistani soldiers. Their warnings of mercenary infiltration supposedly led the Indian army to underestimate the seriousness of the intrusion. Yet, this critique did not explain why intelligence analysts should have predicted the same scenario which had previously appeared irrational to their consumers in the military. Instead, all that it did was use hindsight as a basis for evaluating foresight.

This chapter challenges the still-dominant view that the kargil crisis represented an "intelligence failure". It suggests that the Indian intelligence agencies accurately assessed Pakistani intentions prior to the kargil crisis. Where they went wrong was in predicting the specific form in which these would be enacted. For their part, Indian military officials created an analytical paradigm (or 'model') that reinforced this incorrect prediction. The model assumed that

Pakistani risk assessments would be based on an objective reading of politico-military factors. It did not consider the possibility of Pakistani miscalculation. In this regard, the kargil crisis was similar to other cases of warning failure, such as the 1941 Japanese attack on Pearl harbour and the 1962 Cuban missile crisis.

Backgrounder

Kargil is an administrative district in the Indian portion of kashmir, adjacent to the LoC with Pakistan. It is dominated by steep, barren mountains, whose heights vary between eight and eighteen thousand feet. Indian and Pakistani border posts are situated along the outer ridgeline of these mountains. The posts are poorly served by logistics and are subject to the vagaries of extreme weather during winter. Snowfall and the attendant risk of avalanches make forward patrolling along the LoC risky for both sides. Until 1999, there was an unofficial policy of mutual withdrawal during winter months, when the most isolated outposts would be vacated. They would be reoccupied once the summer thaw set in.

Over 70 percent of the kargil population is shia Muslim. This makes the district unique in Indian-administered Kashmir: it is the only Muslim-majority area which has not been affected by separatist violence. Since kashmiri rebels and their Pakistani patrons are overwhelmingly sunni, the shias of kargil have remained apathetic to them. (such aloofness stems from sectarian tension, which has been worsening in Pakistan since the 1980s.) The district is sparsely populated, with 81,000 inhabitants spread over approximately 14,000 sq km. It has very little vegetation and no forest cover. Unlike the nearby kashmir valley, therefore, kargil is demographically and topographically unsuited for guerrilla warfare. Responsibility for local defence rests with the Indian army's 121 Infantry Brigade, which is headquartered in the district's largest town (also called kargil). The brigade reports to the 3 Infantry Division, which is based in a larger adjoining area known as Leh. Connecting Leh and kargil is a road called national highway 1a. This road is a logistical lifeline for troops deployed in the mountaintops, and is one of only two supply routes that goes on to connect Leh with the rest of India. It runs roughly parallel to the LoC at distances of between six and twelve miles and is, therefore, vulnerable to Pakistani artillery fire. since the highway at kargil is snowed in between september and June every year, uninterrupted use of it is vital to stockpiling of military rations and munitions during summer. The other supply route to Leh, though outside the range of hostile artillery, is longer and is also closed during winter. Accurate and well-timed bombardment of the kargil-Leh road would paralyse Indian fighting capacity at the local level.

The proximate origins of the kargil crisis can be traced to the nuclearisation of South Asia in May 1998. India began the process by conducting five nuclear tests; Pakistan swiftly reciprocated. Both countries were subjected to international sanctions. Both, however, grew more self-confident, convinced that their nuclear arsenals would henceforth protect them from external threats. On the Pakistani side, key thinkers came to believe that with foreign governments concerned about the possibility of nuclear war, Islamabad could ratchet up support to the kashmiri separatist movement. Any risk of Indian military retaliation stood nullified by the likelihood of third-party intervention, should war seem imminent.

In mid-November 1998, four Pakistani Generals, among them the new army Chief Pervez Musharraf, drew up plans for an incursion into Indian- administered kashmir, codenamed Operation Badr. The primary objective of the operation was to interdict vehicular movement along the kargil-Leh road. a secondary aim was to relieve pressure on the kashmiri separatist rebellion, which had recently suffered significant reversals vis-a-vis the Indian state. In order to sustain operations, many insurgent groups had already been compelled to recruit foreign-born mercenaries on two-year contracts. According to an Indian estimate, by 1998, these mercenaries (most from Pakistan and Afghanistan) accounted for 70 percent of all insurgent movement across the LoC. reconnaissance for Operation Badr began in November 1998. Pakistani troops probed Indian defences along the LoC, looking for gaps through which large groups of men could infiltrate. Due to the extremely rugged terrain at kargil, they found several such gaps. Unmanned aerial vehicles were used to verify the laxity of Indian border security. Actual movement of troops into Indian territory, however, did not begin until at least late February 1999. While it was underway, a significant shift was occurring in Indo-Pakistani relations.

Under pressure from the international community, new Delhi and Islamabad initiated a 'peace process' in early 1999. Secret talks culminated in a much-publicised prime ministerial summit, held on 20 February at Lahore in Pakistan. A tentative roadmap towards resolving bilateral disputes was developed, and both governments received accolades for attempting to normalise relations. Despite lack of substantive progress, an illusion of cordiality had been introduced into Indo-Pakistani relations. It was only shattered when the Indian forces discovered the intrusion at kargil, in early May 1999.

By that time, approximately 1,700 Pakistani soldiers had crossed the LoC and occupied seven prominent hilltops overlooking the kargil-Leh highway. They were supported by four times as many logistical troops, who ferried supplies from base camps in Pakistan Occupied kashmir (Pok). The intruders wore civilian clothing, but carried their full complement of military hardware, including assault rifles, machineguns, landmines, mortars and man-portable surface-to-air missiles. They were scattered along 130 defensive positions, located at depths of between three and eight kilometres into Indian territory. Each position was manned by 10 to 20 soldiers. There were also 130 artillery pieces deployed on the Pakistani side of the LoC, ready to provide the intruding forces with covering fire once Indian troops began their anticipated counter-attack.

The kargil intrusion was a historical discontinuity event in two ways. First, never before had Pakistan launched a military offensive against India without a prior build-up of diplomatic tensions. That it would carry out an offensive even as peace talks were underway appeared unthinkable. Second, never in the ten-year history of the kashmir insurgency had Pakistani troops crossed the LoC with the intention of occupying territory on the Indian side. either they skirmished with Indian troops on the LoC itself, or crossed over in mufti to operate as guerrilla fighters in rear areas. Predicting the Kargil intrusion would have required explaining why these two trends in Pakistani strategic behaviour were about to be broken. having outlined the geographic and historical setting of the kargil crisis, this chapter shall now describe the charges that were subsequently levelled against the Indian intelligence agencies.

Intelligence Failures and Intelligence Gaps

Discourse on Indian intelligence has long tended to be consumer-driven, which means that it is shaped by the institutional biases of executive agencies such as the police and military. Such biases often fail to differentiate between strategic and tactical intelligence, presuming instead that all activities loosely labelled 'intelligence' fall under the purview of full-time spy bureaucracies.

Meanwhile, lack of communication from the professional intelligence community, in the form of declassified reports, perpetuates this conceptual misunderstanding.

What results is conflation of 'intelligence failures' with 'intelligence gaps'. Public and political debates assume that lack of any information, no matter how specific, amounts to an intelligence failure. They ignore narrower academic definitions, which are more rigorous and can be tested against historical evidence. One example is the definition coined by Abram Shulsky and Gary schmitt, an intelligence failure is essentially a misunderstanding of the situation that leads a government (or its military forces) to take actions that are inappropriate and counter-productive to its own interests. Whether it is subjectively surprised by what happens is less important than the fact that the government or the military is doing or continues to do the wrong thing.

This definition suggests that mere occurrence of an unforeseen event does not automatically qualify as an intelligence failure. Provided decision-makers have been accurately warned about the dangers of the strategic environment within which they have to operate, even a short-term or tactical surprise can do little damage to national security. This is because such surprises are the inevitable result of gaps in intelligence coverage – a hazard faced by intelligence agencies the world over. Even within the most professional and well-resourced intelligence agencies, trade-offs have to be made whenever intelligence resources are allocated to monitor threats. Coverage tends to be optimised to focus on the most likely, as well as the most severe, threats. In the process, a threat scenario that is neither very likely nor very harmful to national security is bound to get lower priority.

Following kargil, the debate on intelligence performance swiftly took on an accusatory and counter-accusatory character. There were four main agencies relevant to this debate. Foremost was the research and analysis wing (r&aw – India's foreign intelligence agency). It was primarily focused on monitoring Pakistan and China. Most of its personnel were police officials, with a handful recruited from other government departments. Military representation existed mainly in the form of an analysis cell headed by a two-star Genera roughly 25 percent of the agency's output dealt with military intelligence, the remainder being dedicated to political and economic intelligence, next in line was the Intelligence Bureau (IB) – described by some as the oldest intelligence agency in the world. Set up in 1887 to track dissident activity, it evolved into a highly efficient counter-intelligence machine. This agency had the lead role in combating terrorist and insurgent threats to India, irrespective of whether these threats were indigenous or foreign-sponsored. Although theoretically barred from operating overseas, IB operatives posted along India's western borders routinely collected intelligence on Pakistan, owing to Pakistani involvement with several Indian rebel groups.

Third, there was the Joint Intelligence Committee (JIC), an apex body set up to prepare intelligence estimates. It was attended by representatives of all leading intelligence agencies, including the intelligence directorates of the Indian armed forces. Despite its high position in protocol terms, the JIC had been undermined by turf wars. The committee merged with the secretariat of the national security Council secretariat (nsCs) in 1998, only to be reconstituted in 2006 as a separate entity. For the sake of simplicity, this chapter will treat the 1999 nsCs as synonymous with the JIC.

Lastly, there was army intelligence. This term collectively encompassed all Indian army personnel engaged in intelligence work. It included battalion-level intelligence staff, through to the Directorate-General of Military Intelligence (DGMI), which was headquartered in New Delhi. 'Army intelligence' was less an organisation than a system, whereby information flows were streamlined to ensure that time-sensitive data reached local commanders, who needed it most. Unlike the JIC, IB and r&aw, it did not have a discernible hierarchy, but its exclusively military nature marked it out as distinct in the intelligence process.

Between them, r&aw, IB and army intelligence produced 43 reports during the period June 1998-May 1999, which were later found to have a bearing on Pakistani intentions at kargil. another two reports were generated by Indian border guards stationed in the area. Thus, a total of 45 indications (or 'signals') of Operation Badr existed in the Indian intelligence system by May 1999. Of these, 11 reached the JIC. They were hidden among 8,400 other reports ('noise') sent to the committee, which were completely irrelevant to the looming crisis. An analysis of the origins and destinations of the 45 relevant reports is revealing:

l army intelligence produced 22 reports, none of which were shared with any civilian agencies, including the JIC. Data exchanges were confined to local commanders at the brigade and division level, and only three reports were even passed up the chain of command to DGMI Headquarters in Delhi. Later, the Army justified this practice by arguing that intelligence only needs to flow one-way, in the direction of the end-user.

R&AW generated 11 reports, of which seven were widely disseminated and four were shared only with the army. IB produced 10 reports, of which three were distributed widely, five went exclusively to the Army, and two were sent to the Ministry of Defence.

Only two reports, one each from IB and r&aw, were copied to the Prime Minister. The remainder were disseminated through inter-service liaison channels. Any information of tactical importance was shared directly with 121 Infantry Brigade.

The official *Kargil Review Committee Report*, published in 2000, makes it clear that Indian intelligence agencies possessed less information than the army on Pakistani military activities. Whatever information they did have was shared with the army, at levels deemed appropriate to the nature of the information. The army, on the other hand, being both a producer and a consumer of intelligence, did not share information with anyone outside its uniformed chain of command. Despite this, many commentators feel that the IB and r&aw bear the bulk of the blame for not sharing information with the JIC.

The intelligence agencies had predicted heightened Indo-Pakistani tensions in 1999. These tensions were, however, forecast to originate from mercenary infiltration across the LoC, not a localised invasion by Pakistani troops. Building on this point, in 2006, a former Indian army chief explained that the failure to correctly identify the intruders when they were first detected, slowed the Indian military response. The army initially viewed reports of the intrusion without alarm. Only once its casualties began building up did suspicions arise as to the nature of events at kargil.

Differing implications of the terms *infiltration* and *invasion* lie at the centre of criticism made against intelligence warnings about the kargil crisis. All commentators agree that the Indian intelligence agencies did not predict a cross-border offensive by regular Pakistani troops, aimed at seizing territory. Where they disagree is on whether the Indian army can use this to excuse its own laxity in patrolling the LoC, despite warnings of infiltration. Is it reasonable to expect intelligence operatives to deliver a superlative performance when Indian soldiers are themselves complacent about the threat environment? The answer to this question requires examining: just what did the intelligence agencies predict in 1998-99, why did they do so, and what were the consequences of these predictions?

INTELLIGENCE ESTIMATES PRIOR TO THE CRISIS

Before May 1999, Indian intelligence analysts and military commanders were united in believing that Pakistani decision-makers would have to be irrational to provoke a military confrontation. Their assessments stemmed from two different perspectives: r&aw analysts were focused on the political and economic situation within Pakistan, while military officials held that attacks across unguarded portions of the LoC, including kargil, would not be logistically sustainable. Any form of cross-border offensive in such areas, including guerrilla warfare, seemed inconceivable due to the rough terrain.

Assessments of the Pakistani military threat were partly influenced by estimates from r&aw, gauging the likelihood of hostilities. These estimates are routinely produced every six months. The first estimate disseminated after the May 1998 nuclear tests noted that Pakistan was going through an economic crisis, which had worsened following the imposition of international sanctions. Dated 06 October 1998, it observed that lack of funds would adversely affect Pakistani war-fighting capabilities. In its next six-monthly report, circulated on 31 March 1999, r&aw concluded that financial constraints would make Pakistani initiation of a war irrational. It predicted that, to escape pressure from hawks within his government, the Pakistani Prime Minister would escalate support for cross-border infiltration by mercenaries.

The economic factor is central to understanding why intelligence analysts were convinced of the improbability of a military attack. Unlike its much larger Indian counterpart, the Pakistani economy was heavily dependent on foreign aid. Its national debt in 1998-99 was rs. 2,880 billion, as against a Gross national Product (GnP) of rs. 2,760 billion. The government had been forced to borrow rs. 600 million per day (about $10 million), just to meet expenses.

The dollar value of the Pakistani currency had plummeted, leading to a balance of payments crisis. Since most of the equipment used by the Pakistani armed forces was imported, shortage of cash had affected maintenance programmes. This formed part of a trend that pre-dated the nuclear tests; in 1997, the Pakistan air Force Chief had admitted that his service lacked the capacity to undertake offensive action in a future war. his forecast was vindicated during the kargil crisis, when the air Force refused to provide close air support to Pakistani troops fighting on the Indian side of the LoC seen from this perspective, the r&aw was not wrong when it highlighted the inability of Pakistan to undertake large-scale offensive action. The problem was, Operation Badr was not a large-scale offensive. It was confined to a narrow front, and carried out through the use of troops already deployed in- theatre. The guiding logic behind it was an erroneous belief that India had only two retaliatory options, both of which favoured Pakistan. Either the Indian forces would overreact by invading Pakistan, thus, prompting international intervention, or they would mount a desultory counter-attack, which would be easily thwarted.

The plan and its underlying presumptions were not subjected to critical scrutiny, for fear of compromising secrecy. Instead, it was blithely assumed that international pressure would limit the quantum of firepower that India could deploy. Furthermore, occupation of higher ground was expected to convey a tactical advantage to the Pakistani troops in daylight skirmishes. No one seems to have recognised that at night, the advantage would shift to the Indian troops, who could stealthily advance through low-lying terrain hidden in shadow. Certainly, no one envisaged that India might launch a local counter-attack backed by aircraft and heavy artillery, while retaining the option of escalating hostilities further.

Foreknowledge of the Kargil intrusion would have required that Indian intelligence analysts not only know Pakistani plans, but also the logic behind them. Owing to the lack of an elaborate multi-disciplinary collection system in the region, such details were not available. R&AW had maintained a skeletal presence in Kargil since the 1970s, as a consequence of budget cuts. Most data on Pakistani military activities came from technical surveillance, which was sporadic and patchy at best. This information only told analysts *what* the Pakistanis were doing, not *why* they were doing it. Even so, both r&aw and IB picked up some indications of escalating hostile activity, which prove that they were aware of the changed threat environment following the 1998 nuclear tests.

On 02 June 1998, the IB dispatched a note to the Prime Minister, containing details about Pakistani logistics-building efforts along the LoC, in the area opposite Kargil. It predicted that, having acquired a nuclear umbrella, Pakistan was likely to push mercenaries into kargil. The note was personally signed by the then-IB chief – in protocol terms, a sign that its contents were extraordinarily sensitive and warranted follow-up action.

The only other high-level warning was an r&aw note prepared on 26 February 1999, as the Prime Minister returned from the Indo-Pakistani peace summit at Lahore. It stated that elements within the Pakistan army leadership were opposed to rapprochement. Such an assessment would have dispelled any illusions that bilateral tensions were likely to subside in the coming months. Seen with the benefit of hindsight, both warnings, while of a general nature, accurately conveyed the instability of the emerging politico-strategic situation.

Over the winter of 1998-99, r&aw and IB predicted an escalation of mercenary infiltration into Indian-administered Kashmir, with a thrust in the direction of kargil. In its October 1998 threat assessment, r&aw had even warned that the Pakistan army might launch "a limited swift offensive with possible support of alliance partners" – a reference to mercenaries.

Seen against its otherwise sanguine estimate of the risk of war, this statement appeared incongruous and immediately prompted verbal queries from the Indian army. The next six-monthly threat assessment omitted any reference to a "limited offensive" and depicted the Pakistani threat as consisting only of mercenaries. By eliminating the ambiguity inherent in its earlier estimate, r&aw projected a binary perception of the Pakistani threat, which was a mistake on its part.

Credible claims have since been made that r&aw was informally pressured to retreat from the alarming projection it had made in October 1998. According to this theory, the agency's warning of a limited offensive did not sit well with prevailing political narratives. Such narratives stated that the nuclearisation of the subcontinent had made India safer, because it would henceforth deter Pakistani aggression. A contrary assessment, arguing that the overt threat of nuclear war had weakened India's conventional military deterrent, would not have been welcome either in the military or political establishments. By omitting to include a limited war scenario in its March 1999 assessment, r&aw might have capitulated to the wilful blindness of its consumers.

Thus far, this chapter has described the criticisms that were made against the Indian intelligence agencies in the aftermath of the kargil crisis. It has argued that r&aw was focused on assessing threats at the grand-strategic level, and did not devote equal attention to lower-order threats which Pakistan had the economic and military means to pose. The next section shall describe how intelligence forecasts about mercenary infiltration matched threat assessments by military commanders at kargil.

Threat Assessments and the 'Afghan Model'

The headquarters staff of 121 Infantry Brigade and 3 Infantry Division gave tremendous cognisance to the implicit difference between *infiltration* and *invasion*. However, they were more alert to the former threat than the latter. Intelligence warnings about mercenary infiltration were followed up by periodic patrols of likely infiltration routes. These were assessed to be in low-lying terrain, such as ravines and dry river beds. Little effort was made to tighten surveillance on the high ridges near the LoC itself. It was on these ridges that the Pakistani intruders built up their presence.

During early 1999, the intelligence agencies did not notice the absence of a crucial indicator, which was integral to their hypothesis about mercenary infiltration. Unlike elsewhere in Indian-administered Kashmir, Pakistani covert operatives had not engaged in preparatory subversion of the local populace within kargil. From all the expectations about infiltration, one would also have expected to uncover efforts to create a supportive infrastructure for the mercenaries in populated areas. However, no evidence of a subversive network was detected in kargil throughout the period preceding the crisis. The sole exception was a 25-person spy ring in an adjacent area, which was dismantled by Indian counter-intelligence in June 1999.

For their part, military officials were convinced that an incursion into kargil by the Pakistan army was a near-impossibility. They believed that it would prove unsustainable in the long term and expected their Pakistani counterparts to recognise that, although logistics networks on both sides of the LoC were rudimentary, India had a relative advantage. During summer, the Indian road network opened for traffic a month earlier than the Pakistani one, and was also capable of carrying heavier loads. Troops and heavy weapons could be rushed to into the area faster than the Pakistanis could be expected to strengthen any newly-occupied positions. If an offensive was to take place in this area, Indian planners reasoned, it would only make sense if it were initiated by them, not Pakistan.

Thus, in February 1999, a wargame categorically ruled out the possibility of a cross-border incursion by the Pakistani military. The Indian officer who suggested this scenario was ridiculed; an experience similar to that of professional intelligence analysts who predict that the enemy might behave irrationally, only to end up themselves being branded as irrational.

Following on from its institutional logic, the Army was quite prepared to believe reports from the intelligence agencies that the primary threat in kargil was one of mercenary infiltration. Given the terrain and its unsuitability for large-scale conventional warfare, one can reasonably surmise that the army might have been far more sceptical had intelligence estimates warned of a cross-border attack. Its commanders only saw what they expected to see. Where did this subjective assessment of Pakistani rationality come from? Why did Indian intelligence analysts and military officials assume that they understood the logic which guided Pakistani decisions? An answer might be found in the analytical model which Indian decision-makers had come to rely on while interpreting Pakistani strategic behaviour.

During the 1990s, a view developed within the Indian military and intelligence communities that Pakistan hoped to re-enact the soviet-afghan war in kashmir. It was widely known that the Pakistani Inter-services Intelligence (ISI) had played a key role in organising the afghan resistance against the Soviet Army. In the words of a former ISI officer, his country had waged a "war of a thousand cuts" against the Soviets and their afghan allies. Attritional attacks to sap troop morale were combined with sporadic daredevil raids on transportation infrastructure and logistical chokepoints.

These tactics, examined collectively, constituted what might loosely be called the ISI's 'afghan model'. As rebellion broke out in Indian kashmir during the late 1980s, senior Indian military officials wargamed the likelihood of covert Pakistani intervention.

They had before them the 'afghan model', which suggested that Pakistan was prepared to support cross-border insurgencies, so long as it could maintain plausible deniability and escape military retaliation. At the time, Islamabad's behaviour rigidly conformed to this paradigm: since 1979, Pakistani decision- makers had demonstrated a strong instinct for self-preservation, and desisted from any actions which might provoke a soviet invasion of their country.

This ability to balance opportunism with prudence was assessed as a constant trait of the Pakistani leadership.

From ground-level studies by Army officers emerged "Operation Topac", a predictive assessment published in July 1989 by the *Indian Defence Review*. "Operation Topac" became a quasi-official interpretative framework used by intelligence analysts to predict Pakistani moves in kashmir. It hypothesised that the kashmir rebellion would consist of three phases, all orchestrated by Islamabad. First, there would be a subversive campaign, aimed at radicalising the kashmiri population and preparing it for revolt. Second, Islamist mercenaries who had fought in the Soviet-Afghan War would be infiltrated into Indian-administered kashmir. Their job would be to attack military bases and block strategic roads at chokepoints. Given the topography of kashmir, there were only two such chokepoints, one of which was kargil. Finally, at an opportune time, the Pakistan army would cross the LoC in support of the kashmiri rebellion, but not before the Indian army was substantially weakened.

"Operation Topac" assumed that the Pakistan army and ISI would calibrate insurgent violence to avoid provoking military retaliation. This assumption was largely vindicated by events: during the 1990s, the ISI observed a degree of restraint while supporting the kashmiri rebellion. For instance, it did not provide the insurgents with surface-to-air missiles, having been warned by r&aw that if an Indian aircraft was shot down, the result would be war. However, like in Afghanistan, the ISI and Pakistan army were prone to occasional miscalculation, which led to these restraints being loosened in 1999. Viewed in historical terms, the kargil intrusion bears a strong similarity to a now-forgotten military fiasco that took place in Afghanistan in spring 1989. With soviet forces having withdrawn from the country, elements in the Pakistani government demanded that Afghan resistance fighters immediately seize power. Under pressure from the ISI, the insurgents launched a frontal assault upon the large garrison town of Jalalabad, close to the afghan-Pakistan border. The attack got bogged down in well-prepared defences that were supported by aerial and artillery bombardment. Of 6,000 insurgents who participated in the assault, 3,000 were killed or wounded. The garrison did not fall. Analysing the episode years later, an ISI officer bitterly noted that the Pakistanis had blundered by moving too swiftly from guerrilla tactics to positional warfare. Once the Afghan rebels presented a fixed target for government forces, they were decimated through superior firepower. As the next section shall show, this is analogous to what happened in kargil a decade later.

An Incorrect Prediction, but a Correct Assessment

The use of denial and deception was an integral component of the Pakistani operational plan at kargil. It aimed to convince the Indians that the intruding forces were mercenaries. The Pakistani planners hoped that this would slow down the Indian response and also deflect international criticism of Islamabad. Initially, these hopes were well-founded. Indian commanders, relying on signal intercepts, came to the conclusion that the intruders were Pashtu-speaking afghans. Operating from past experience, they assumed that a show of force would suffice to send the intruders scurrying back across the LoC into Pakistan. (hitherto, mercenaries in kashmir had shown no hesitation in escaping from a firefight if offered a way out.) Meanwhile, neutral governments decried the escalating tensions along the LoC and avoided naming Pakistan as the aggressor.

On the ground, Indian battalion commanders were ordered, with characteristic bravado, to throw out the intruders. Assaults were mounted in broad daylight, in the expectation that resistance would be slight. As casualties mounted, however, it became clear that unlike past engagements, this time the 'mercenaries' aimed to hold on to territory. By 25 May 1999, the military situation had reached a stalemate. Aerial reconnaissance confirmed that the intrusion had made deep ingress into Indian territory and that the intruders were receiving full logistical support from Pakistan. Over fifty Indian soldiers had died in suicidal assaults against Pakistani defences, and not a single hilltop had been retaken. At this juncture, the Indian government broke with past practice and sanctioned the use of air power and heavy artillery against the intruders.

It had, thus, far treated the situation in kargil as a counter-insurgency operation, with limits on the amount of force that could be used. Battalions deployed to fight in Kargil had, therefore, been ordered to leave their support weapons at base camp, before setting off to fight in the mountains. However, as suspicions built up that the intruders were regular Pakistani soldiers, operational restraints began to be loosened. On 26 May, the Indian air Force commenced bombing missions against the hilltops occupied by the intruders. The same day, the r&aw intercepted a telephone conversation between the Pakistan army Chief and one of his deputies. The intercept confirmed that the intruders were regular Pakistani troops, who were being portrayed as freelance mercenaries by Islamabad in order to escape international censure.

From 26 May onwards, the counter-attack in kargil took on the nature of a limited war. This dramatic escalation had not been foreseen by the Pakistani officers who had planned the intrusion. Over the following weeks, a combination of aerial reconnaissance and artillery bombardment turned the military situation in India's favour. Approximately 150 reconnaissance flights by the air Force produced several thousand photographs of Pakistani defences. Working with these, Indian infantrymen switched from high-visibility to low- visibility tactics, infiltrating past Pakistani defences at night and attacking from the flank. The Army deployed 300 artillery pieces in the combat zone, firing a total of 250,000 shells. Wireless intercepts recorded the desperation which set in among the intruders, with one telling his commanders in Pakistan, "hell has fallen on us." Urgent requests for resupply and reinforcement were overheard.

The help never came, due to inherent flaws in the Pakistani operational plan. Operation Badr had been envisaged as an infantry battle, because that was where Indian and Pakistani forces were most evenly matched.

On other counts, be it air power, naval power or armoured strength, India had a strong advantage. Faced with these factors, the Pakistani planners had convinced themselves that hostilities would be contained. They believed that the Indian army was heavily committed in counter-insurgency duties and could not spare troops for an attack on Pakistan. They ignored the fact that by insisting that the intruders were freelancers, Islamabad deprived itself of a *locus standi* to overtly intervene on their behalf. India could escalate hostilities locally as much as it pleased, while Pakistan could not reciprocate without admitting its involvement in the intrusion. Furthermore, by committing the intruders to hold on to their territorial gains, the Pakistani high command had in effect, gifted the Indian Army a fixed target against which to concentrate overwhelming firepower. The premature shift from insurgency to positional warfare at kargil, as at Jalalabad in 1989, had played to the strengths of counter-insurgent forces.

These realities did not stop some Pakistani Generals, both before and during the crisis (and in some cases, even after) from indulging in delusions. One of the planners of Operation Badr asserted in a secret meeting in May 1999 that Pakistani troops would soon march into Indian-administered kashmir, "to mop up the bodies of hundreds of Indians left hungry, out in the cold."

Others conjured up visions of the '*Mujahideen*' (as Islamist mercenaries were reverentially called in Pakistan), conquering Kashmir. Later, they insisted that had the United states not brokered a Pakistani withdrawal, the Indian garrison in Leh would have surrendered to the kargil intruders.

From these statements, it appears that years of Islamist propaganda (officially promoted within Pakistan), had partly warped the professionalism of the Pakistani officer corps. According to the well-known Pakistani commentator shuja nawaz, over the two decades preceding the kargil intrusion, several top officers had been blinded by religious fanaticism. Military planning was punctuated with sentiments like "by the Grace of God, we will put 10,000 rounds over there and *Inshallah* the enemy will be routed." Plans were not subjected to detailed logistical analysis because these officers felt that "you cannot quantify God's Grace." reportedly, at least two of the Generals who planned Operation Badr were Islamist radicals. If true, this would explain why they were so confident that supply constraints and Indian counter-attacks would not dislodge the intruding forces. As events turned out, the kargil crisis not only played to India's strength in heavy firepower, but in political terms, it was spectacularly ill-timed. Coming just months after the Indo-Pakistani peace summit at Lahore, it left Islamabad internationally isolated. Whereas a year previously, India had been criticised for initiating the nuclearisation of south Asia, new Delhi was suddenly being lauded for its restraint in not invading Pakistan. Pakistan, on the other hand, went from enjoying sympathy over its economic plight to being regarded as a nuclear-armed rogue state. As one Indian writer noted, the kargil crisis accomplished what Indian diplomats had long wanted: to expose Pakistani involvement in the Kashmir rebellion. He commented, "it is not sufficient if Pakistan is a 'rogue' covertly, it must be seen to be one." From this perspective, the Kargil crisis proved beneficial to India.

The Challenge of Estimating 'Rationality'

The kargil crisis shares one common feature with at least three other cases of warning failure. These are Pearl harbour, the Cuban missile crisis and the Yom kippur war. In all these cases, early intelligence estimates were grounded in assumptions that were later found to be flawed. The 'mindset' used by analysts to anticipate hostile action was different from that used by the adversary. However, it closely reflected the mindset of intelligence consumers – leading by default to an analytical failing known as 'mirror-imaging'.

Traditionally, 'mirror-imaging' has been perceived as emanating from a tendency to project one's own belief system onto the adversary. Many intelligence veterans, however, insist that they are aware of the need to avoid it. From the case of kargil, it appears that mirror-imaging can creep into warning analysis through inputs from intelligence consumers. When strong- minded consumers, such as military officials, develop fixed ideas about how the enemy will behave, intelligence analysts might end up internalising these ideas. The result is that intelligence producers fail to warn of threats which their consumers do not believe exist.

For instance, in 1941, Us naval commanders believed that Pearl harbour was so well-guarded as to be immune from attack. They viewed Japan as a presumptuous midget that dare not launch a strike against American interests. Japan, after all, had one-tenth the industrial capacity of the United states. Its ability to survive a long war was doubtful and its ability to win one, non-existent. Given that the Us had shut off Japan's oil supply in July 1941, it seemed reasonable to expect that Tokyo would soon negotiate on American terms, rather than self-destruct. As tensions escalated, the possibility that Japan might attempt to obtain oil by conquering Southeast Asia was factored into intelligence estimates. What was not predicted was that it would begin its campaign by aiming to paralyse the one naval force which could interfere with its plans: the US Pacific Fleet. Japanese officers who planned the attack had a different mindset from their American counterparts. They believed, naively, that after a flurry of retaliatory action, the Us would write off its losses and reconcile to Japanese dominance over the eastern Pacific.

In september 1962, Us intelligence analysts estimated that soviet leader Nikita Khrushchev would not deploy ballistic missiles in Cuba. Supporting their argument was the fact that the soviet Union had never deployed such missiles outside its own territory, and so was hardly likely to deploy them in a country just under kilometres from the United states. To do so would seriously undermine Us national security and provoke Washington into drastic action: a scenario that khrushchev was expected to anticipate. Although subsequently proven wrong by events, this estimate at the time, mirrored conventional wisdom within the Us government and academia. Only much later did Us analysts learn that khrushchev feared a coup by hardliners in his government more than he feared the likely Us reaction, and so felt compelled to engage in brinkmanship over Cuba.

Lastly, prior to the Yom kippur war, Israeli military intelligence had assessed arab military threats according a paradigm known simply as 'The Concept' (*Ha-konceptzia*). It contained two assumptions. First, Egypt would not attack Israel unless it first acquired Soviet bombers and surface- to-surface missiles, to neutralise the threat that Israeli airpower posed to egyptian tanks. Second, Syria would not attack Israel without the active support of Egypt. Although the second assumption has thus far remained valid, the first ceased to have effect a year before the Yom Kippur war broke out. During the following twelve months, egyptian decision-makers started to believe that they did not need to attack Israeli airbases with bombers and missiles in order to neutralise enemy airpower. All they needed was a surface-to-air missile shield behind which their armoured columns could advance into Israeli territory. This change in mindset was not detected by Israeli intelligence analysts. The latter continued to believe that unless arab regimes were certain of decisive military victory, they would not gamble on their own domestic stability by starting a war.

The adversary's actions defied conventional wisdom as to what would constitute 'rational' behaviour. Risk-taking in a low-returns/ high-costs context is widely perceived to be an aberration in international politics. Hence, Indian analysts believed in 1999 that the Pakistan army would not initiate hostilities which it could not control, especially in a remote and desolated area. Their assumptions, even if specifically wrong, were eventually proved vaguely right by events. Although allegations of 'intelligence failure' will continue to be made, a fair evaluation of kargil would echo the observation of robert Jervis, "states rarely expect their adversaries to behave foolishly. The Pakistani incursion simply did not make sense; it was therefore sensible of India not to expect it." Kargil suggests that predictive models which hinge around an inherently subjective estimate of adversary rationality are vulnerable to being discredited by events, sooner or later. On the one hand, the Indians in 1999 made reasonable assumptions that Pakistani strategic behaviour would continue as before, with low-level provocations growing more frequent. These assumptions were grounded in trend analysis and military logic, and appeared to be supported by secret intelligence. On the other hand, the same assumptions also led the Indians to arbitrarily (and wrongly) conclude that the Pakistani military leadership was permanently disinclined to take a gamble and escalate hostilities to qualitatively new levels. In this, Indian intelligence analysts and military commanders were projecting their own behavioural codes onto the adversary and using them as an inferential short-cut for predictive analysis. By focussing on the military balance between India and Pakistan, they overlooked the temptations that a surprise attack would offer to the side that was manifestly weaker in overall terms.

A Decade Later, Little has Changed

In the years since kargil, considerable effort has gone into remedying shortfalls in intelligence coverage through enhanced technical surveillance in forward areas. Communications interception is better and the increased use of satellite imagery and Unmanned aerial vehicles (Uavs) could well provide a battlefield advantage to the Indian forces in future conflicts. At the national level, efforts have been launched to implement comprehensive intelligence reform. These have only been partially successful, due to the inherent challenges of overcoming bureaucratic hurdles. The Ministry of Finance, for instance, has reportedly been reluctant to release funds for the salaries of additional personnel recruited by the IB. Meanwhile, partisan politics has reduced the effectiveness of liaison between the state police forces and the IB, despite the creation of task forces intended to synergise domestic intelligence operations.

Although it is desirable that the planned upgradation of intelligence collection and liaison arrangements be followed through, another problem does not seem to have been addressed at all. Intelligence analysis and assessment remains an overlooked area. This chapter has argued that the central issue in 1999 was not poor information collection or sharing. Rather, it was the inability of analysts to provide a convincing explanation for why Pakistan would behave in a counter-instinctive manner and start a conflict when it was relatively weak. Without an organising paradigm to interpret raw information correctly, whatever data was available was used to fill out a predictive model that ultimately proved misleading. Almost a decade later, the Mumbai terrorist attacks of November 2008 suggested that this analytical blindspot remains relevant to the quality of intelligence warning.

One of the reasons why intelligence agencies can be slow to detect the emergence of new threats is 'discourse failure'. This is a phenomenon which originates outside bureaucracies, in the world of academic and public debate. Basically, when the terms of discussion around an issue are limited by politically correct guidelines of what can or cannot be said, intelligence analysts find themselves unable to warn of outlier threats, i.e. wildcard scenarios. Us intelligence agencies, for instance, avoided referring to terrorism as a religiously-defined problem during the 1990s, for fear that such a conception could be criticised for being racist or Islamophobic. Due to institutional self-censorship, analysts were unable to officially express concern about the possibility of an attack by *jihadists* from Saudi Arabia, a key Us ally. Perhaps a similar problem hobbled Indian agencies during 2007-08, in a political climate wherein the emphasis was on normalising relations with Pakistan through 'people-to-people contact'. The hype built up around cultural diplomacy might, in fact, have inadvertently restricted the agencies' freedom of expression, by moulding political discourse in a manner that precluded frank discussion of pessimistic views.

Like kargil, there were plenty of misleading signs in 2008, indicating a different kind of threat from the one that was actually brewing. The trajectory of *jihadist* violence just prior to the Mumbai attacks pointed towards growing domestic radicalisation, in the form of the Indian Mujahideen. Pakistan was experiencing political turmoil. A theory, thus, appeared within the intelligence establishment that Islamabad was not in full control of terrorist groups and could not be blamed for every attack on Indian soil. Given that Benazir Bhutto had recently been assassinated by unknown elements, this argument could not be dismissed as pure nonsense. What it overlooked, however, was the possibility that

Pakistani officials might seek to regain control over *jihadist* groups by facilitating a major offensive against India. From the 2010 interrogation of Lashkar-e-Tayyeba operative David headley, it is now known that mid-ranking ISI trainers conceptualised the Mumbai attacks as a way of preventing a split in the Lashkar. To prevent elements within the group from turning against the Pakistani state, they encouraged it to attack India. Although this probably seemed a rational decision to them, it nearly provoked a war that could have turned nuclear and wiped Pakistan off the map.

There is no indication that prior to the event, either the r&aw or IB had any inkling of the logic that eventually led the ISI to plan and support the Mumbai attacks. This begs the question: since Kargil, has anything being done to strengthen the rigour of intelligence analysis, so that Pakistani decision- making can be wargamed? Equipment upgrades, though essential in detecting preparations for a conventional or sub-conventional attack, will not by themselves identify the thought processes behind Pakistani adventurism. For that to happen, thinking patterns within the Indian security and intelligence communities need to be modified slightly.

To start with, multi-level joint training between the analytical components of civilian and military agencies could introduce greater harmony at the interface of strategic and tactical intelligence. For surprises like kargil and Mumbai to be avoided, it is essential that decision-makers in the military understand the work ethos of the IB and r&aw. Both these agencies serve a range of consumers, with the information needs of the Prime Minister being their highest priority. Producing a briefing note for the political executive will always come before producing a situation report for the army – that is the nature of the business. Once military commanders realise this, they will be less quick to accuse the agencies of intelligence failure whenever there are knowledge gaps in tactical reporting. It is worth remembering that even during the 1971 war, when the army praised the r&aw for a brilliant strategic-level performance, there were complaints about the quality of its tactical intelligence. Another advantage of closer interaction between civilian and military intelligence cadres would be a transformation of their respective work cultures. The IB and r&aw could be militarised to some degree, while the armed forces intelligence directorates could be professionalised. Typically, civilian agencies indoctrinate analysts to believe that their job is "telling truth to power." The purpose of intelligence, as they see it, is to convey the complexity of the international environment to decision- makers in as nuanced and objective a manner as possible. Towards this end, they avoid making categorical judgments and hedge their bets while providing an assessment. Such conservatism is integral to their work culture; civilian intelligence analysts are not trained, or meant, to make snap decisions. Military analysts, on the other hand, constantly seek to convert ground-level knowledge into a blueprint for operations. What they seek is 'actionable' intelligence; something civilian agencies are not institutionally attuned towards. Bringing both types of analysts together in training sessions and simulation exercises might boost intelligence coordination to a greater extent than routine information-sharing forums could.

UNPACKING INDIA'S INTELLIGENCE AGENCIES

India's premier civilian intelligence agencies are the Intelligence Bureau (IB) and the Research and Analysis Wing (R&AW): the former deals with domestic intelligence, the latter with foreign intelligence. In addition, there are the intelligence directorates of the Armed Forces. Upon independence in 1947, the IB was the only real intelligence organization that India possessed. Set up in 1887, this agency had been responsible for political surveillance during the British Raj. After India achieved independence, it was entrusted with additional responsibility for foreign intelligence. Between 1951 and 1968, the agency struggled to meet a very broad mandate in a deteriorating regional security environment, as India's relations with its neighbours plummeted. China and Pakistan launched unexpectedly audacious attacks in 1962 and 1965 respectively, prompting a realization among the Indian political leadership that a specialized agency was needed for foreign intelligence. Thus, in 1968, the R&AW was created. Both agencies have had impressive successes and some rather nasty intelligence failures. The IB in particular is considered as one of the most professional intelligence agencies anywhere in the world, and has achieved a remarkable measure of success in counterterrorism.

Logically it would. In practice, however, the West wishes to retain a modicum of goodwill from the Pakistani Inter-Services Intelligence (ISI), which alongside the Pakistani army, is the main sponsor of terrorist groups in South Asia. This is to be expected, since the ISI is rationing out counterterrorism cooperation to those states that avoid criticizing Pakistan too harshly for its Janus-faced policy of supporting some terrorists while fighting others. With its large and restive Pakistani-origin population the United Kingdom in particular, is vulnerable to Islamabad's blackmail in maintaining homeland security. British security officials privately admit this. Even so, the West's intelligence partnership with Pakistan is a tactical affair borne not of mutual trust and goodwill, but out of the negative leverage that Islamabad retains through its dalliance with radical Islamist groups. As such, Indian intelligence agencies share a strategic interest with their Western counterparts in dealing with the threat of Pakistan-sponsored terrorism that transcends the divide-and-conquer games that the ISI tries to play. There is already an informal intelligence-sharing network where security agencies compare notes on the kind of dissimulation efforts that Pakistan engages in. Also, India's intelligence services make reliable long-term partners for the West since they operate under full civilian control and within a democratic framework.

The general consensus in India is that Zawahiri's call is part of an effort by the ISI to regain plausible deniability for its sponsorship of terrorist attacks, which was partially lost after the Mumbai 2008 attack. Knowing that Indian political and public opinion would push for strong retaliation in the event of a future terrorist onslaught from Pakistani territory, the ISI is trying to preemptively deny responsibility by floating the idea that 'non-state actors' are responsible. It has used the same tactic in the past, when in 2007 it created the so-called 'Indian Mujahideen' group, using a small number of jihadists locally recruited from within India to carry out urban bombings. Even if one assumes, however, that Zawahiri's call for jihad against India was made independently of ISI prompting, the fact remains that the core leadership of Al Qaeda is increasingly dominated by Pakistani nationals. A decade-plus of Western counterterrorist operations has badly weakened the Al Qaeda command structure in Pakistan, making it vulnerable to a creeping takeover by local jihadists. So there is no way that Pakistani involvement in a future terrorist attack on India cannot be traced back to that country's borders. What remains an open question is whether the ISI itself can escape international condemnation, which it is hoping to do.

The Indian security establishment has a well-developed doctrine to combat political subversion, and counterinsurgency and counterterrorism are subsets of a larger effort to preserve the democratic framework of the country. This means that working together with communities that are potentially marginalized or disaffected is essential. India's most powerful defense against terrorism has been the consistent patriotism of its Muslim population, which largely recognizes that the ISI is keen to falsely implicate them in cross-border attacks so as to provoke a larger conflict within the country. During the 2008 Mumbai attack for instance, the farcical manner in which Pakistani terrorists tried to masquerade as locals under the moniker of 'Deccan Mujahideen' showed how amateurish the ISI's false-flag operations can be. Indian Muslims are wary of the larger design behind such antics.

Human intelligence from deep-penetration agents, mostly operated by the IB, has been one of the reasons why foreign jihadist groups like Lashkar-e-Taiba have been unable to make any headway in Kashmir over two decades, despite the terrain favouring guerrilla warfare. Likewise, resistance from mainstream Islamic seminaries in India has been a barrier to Pakistani subversion. A classic example was the May 2008 fatwa against terrorism issued by the Dar-ul Uloom seminary in Deoband, the most prestigious in South Asia. Pakistani Islamists were greatly upset by this development, and also by the manner in which the seminary called upon Indian Muslims to show complete solidarity with their Hindu countrymen after the terrorist attack on Mumbai.

Neither the IB nor the R&AW is a law enforcement agency. They need to cooperate with Indian police forces as a matter of routine. This is helped by the fact that both agencies are dominated by career police officials. In particular, the IB draws its senior cadres exclusively from the Indian Police Service (IPS) which makes for a strong sense of camaraderie. Even so, competition between various state police forces, or within a police force, is an unfortunate reality and this has in the past, led to some counterterrorist operations being bungled by the implementing agency. Generally, the focus is on preventing terrorist attacks in large cities, where the police have extensive local intelligence networks based on human sources. The IB and R&AW possess significant technical interception capabilities which can be used to generate strategic intelligence on Pakistani and other foreign threats. However, while coverage might be higher in some regions rather than others, this has to do with the inevitable need to optimize resource allocation towards the most pressing concerns. The fact that numerous Pakistani spies and terrorists have been arrested in remote rural areas, often close to international borders, demonstrates that coverage can be quite extensive.

That there are competing priorities among different ministries is only to be expected in a country with a unique policy problem: how to ensure equitable and sustained economic growth by integrating with the global economy, while at the same time acquiring the structural integrity that strong Western states have in policing their borders. Indian intelligence agencies have been badly hit by economizing trends in the past, and this has usually resulted in information gaps. Projects to streamline data collation and storage have been stymied by the reluctance of some agencies to share information which their counterparts do not need for operational purposes. While I would agree that bureaucratic rivalries are unhelpful, it is also worth remembering that even supposedly efficient intelligence services such as those in Israel have suffered from turf warfare (known colloquially as 'The War of the Jews'). In Northern Ireland, British intelligence agencies have competed for local sources, going to the extent of undermining national counterterrorist efforts in the pursuit of narrow organizational interests. And prior to 9/11, the US intelligence community was no different – counterterrorism coordination meetings were intended for show rather than substance.

6

Purulia Arms Drop and Intelligence Agencies

The recent revelations on the Times Now news channel where the two main protagonists Kim Davy or Neils Christian Neilsen and Peter Bleach of the infamous December 17, 1995 Purulia arms dropping case appeared and answered some questions has the country agog with a sense of disbelief. There are too many unanswered questions and some very uncomfortable and rather dangerous conclusions. I will not go into the details of the case as I presume that the readers must be knowing as to what happened. It is the questions that are still hanging in the air that concern me.

Let us presume that the Indian government had colluded with MI5 and planned these operations to destablize the West Bengal government. Then why did the plane land in Karachi first and then enter the Indian airspace? Why were the services of the Shaheen Airlines taken for the AN29? Shaheen is a known ISI front. Who financed the whole operation? There are names of two Hong Kong based businessmen doing the rounds. If it was an Indian government sponsored operation in which RAW and MI5 were involved then why did they need to have foreign financiers? If we go by the reports in The Independent it seems that Kim Davy financed the whole operation with money raised from smuggling gold etc. If so why have letter rogatories sent to Hong Kong for the two men alleged to have been behind this?

Once the arms had been para-dropped and made news headlines and the plane had reached Bangkok why did they chose to come back into the Indian airspace? It was clear that after the furor in the media things would be difficult for the men who had undertaken this operation. And if it was indeed an Indian government sponsored operation why did they force land the plane in Mumbai? They could have still whisked it away. How did Kim Davy escape once the plane had landed in Mumbai? If it was indeed Pappu Yadav and J.K. Dutt who facilitated Davy's escape is it not true that there must have been forces higher up in the bureaucracy who had given this a go ahead. Who are these men?

Where is the 30 page confession that Peter Bleach gave Indian authorities most notably J.K. Dutt and why was it not that made public? It was in the media that Peter Bleach and the Latvians were given presidential pardon because of pressure from British and Russian governments. But it now appears that Mr. Bleach was given a pardon in a hurry so that he does not depose before the High Court judge of Kolkota. What was it that Peter Bleach was to reveal that was so damaging that Indian authorities found it convenient to let him go rather than allow him to speak. It must also be noted that whenever Bleach tried to speak to the media while on the way to the court he was whisked away. Readers may be wondering as to why I have not mentioned or asked anything about the shutting off of the radars which allowed the AN29 to fly into India from Karachi. Well, if it was a MI5-RAW operation then these things can be arranged. But I am sure this must have been after approval from highest authorities of the land. And to say that Pappu Yadav has clout enough to ensure safe passage to Davy is rather farfetched.

If indeed it was a MI5 okayed operation then why did they tip off the Indian authorities about the impending arms drop? They should have been working in tandem with RAW and IB.

From the interview of Kim Davy it was clear that he has strong views about the atrocities committed by the CPI(M) goondas, as he calls it. The man has a history of criminal activity. But he also seems to be a man with a mission. How did this small time criminal turned activist get access to MI5 and RAW? Who got him into the loop of such a massive operation purportedly to destabilize the West Bengal government? Is it normal for intelligence agencies to use the services of dubious characters to further their political ends? The CPI(M) has alleged that arms have also been dropped in Tripura in the past where they are in power. It is also intriguing that for 16 years the Indian intelligence agencies had been telling the courts that they did not know where Kim Davy was while he was giving interviews in Denmark. Also why have they woken up now after all these years and asking for his extradition? What has changed in this time?

The most unnerving question is however about the ISI's role in this operation. It seems that J.K. Dutt was entrusted with 'cleaning up' after the whole operation had come a cropper. Shaheen Airlines had a contract with a company floated by Kim Davy. The plane landed in Karachi and the crew waited for a few days before they were told that a 2 day window was open for them to make the drop. Was the ISI in the know that Indian radars would be shut off to facilitate this operation? Could not the Pakistan government have taken advantage of the fact and launched a military operation? J.K. Dutt is the same man who was the head of the National Security Guards when the 26/11 Mumbai attacks took place. It must also be remembered that generally the head of the NSG is a military officer.

J.K. Dutt is a civilian with intelligence background. Why was he made the NSG chief? It must also be remembered that NSG was finding it difficult to get a plane to Mumbai, or so we are told and arrived only 48 hours after the Mumbai attacks began. Barkha Dutt while interviewing J.K. Dutt had asked whether the handlers had talked to him about swapping the one terrorist we had captured, Ajmal Amir Kasab, for some hostages and he had said that he had not talked to the handlers. He was not very convincing when he said that he had no contact with the handlers of the 26/11 terrorist handlers. Were the Mumbai attacks also had an Indian angle to it. The terrorists both in the Trident and the Taj were never short of ammunition. They could not have carried so much ammunition in their ruck sack. Were there terrorist accomplices already staying in theseHOTELS AS guests and had stocked arms and ammunition for the party from Karachi to arrive? Peter Bleach did caustically remark how well J.K. Dutt had handled the operations at the Trident. It is not uncommon for intelligence agencies of the worst of enemies to be in touch and to exchange notes. If the impending Purulia arms drop was known to the ISI could it be possible that the 26/11 attacks were also known to some in the Indian intelligence and police. The Mumbai police chief Gafoor was sidelined after the 26/11 attacks. I will not be surprised if there were people in the know here in India about the Mumbai attacks well in advance. J.K. Dutt and his colleagues need to answer a lot of questions.

Lastly, it is intriguing that Peter Bleach was given presidential pardon during the Vajpayee government. The BJP representative refused to be a part of the debate on the Purulia arms drop on Time Now. Why was the BJP government so keen to keep things under wraps? It was the Congress led government that had planned and executed the Purulia arms drop, then why did BJP led NDA government play along and let Bleach go.

There are way too many questions. It is a matter of concern that intelligence agencies play games of intrigue and deceit with each other and with the people they swear to protect. The political class and the intelligence agencies need to clean up the mess. Games can be played, but not at the cost of national security and the lives of ordinary citizens.

A word about Kim Davy before I pen down. The man mentioned the organization PUCL (People's Union for Civil Liberties) a couple of times. The man seems to have very strong views regarding the atrocities against the poor by the CPI(M). PUCL is the same organization to which Binayak Sen and Kobad Gadhi belong to.

It is a front for the Maoists in India. Such huge arms cache that included AK-47 rifles, rocket launchers, grenades, night vision glasses etc. can only be used by the Maoists in India. Could Kim Davy be working for the Maoists? The Center has been propping up the Maoists to counter the CPI(M). If at all the CPI(M) government is overthrown this time it will be because of the Maoists alias PUCL. They also have had support from the ISI. Does the clock tick now? Does New Delhi want Davy back because they want to put a leash on the Maoists now that the elections in

Bengal are over and Trinamool and Congress reckon they will be in power. These men were pawns in the hands of the power that be. And power is what politicians are after.

PURULIA ARMS-DROP ACCUSED SAYS OPERATION WAS ORGANISED TO BRING DOWN LEFT FRONT

Niels Christian Nielsen, a Danish national fighting extradition to India for his alleged role in the 1997 air-drop of ammunition over Purulia, has said the operation was sanctioned by the Central government to destabilise West Bengal's Left Front government.

In an interview to *Times Now* television, Mr. Nielsen claimed that the Purulia arms drop was organised as a form of defence against alleged oppression by the CPI(M) in West Bengal. The government of the former Prime Minister, P.V. Narasimha Rao, he said, "saw and approved the plans to arm the defence of these innocent people."

Central Bureau of Investigation detectives have been seeking the extradition of Mr. Nielsen, also known by the alias Kim Davy, saying he was the key figure in organising the drop of several hundred assault rifles, ammunition and explosives to help the murky Ananda Marga cult start an anti-communist insurgency.

Peter Bleach, a former British special forces officer who arranged for the purchase of the aircraft used in the operation, was sentenced to life imprisonment in 2000, along with the Latvian crew. The convicts, however, all later received presidential pardon. In a separate interview to Times Now, Mr. Bleach backed elements of Mr. Nielson's case, saying "the whole objective, my understanding was, to destabilise the Government of West Bengal so that President's rule could be declared in terms of the Constitution [sic.] and the State would have been ruled directly from Delhi."

Little is known about the 1961-born Nielsen, who is reported to be wanted in several countries on charges related to arms trafficking andMONEY laundering. For years a member of the cult, Mr. Davy has also been linked to the South Sudan insurgent leader John Garang.

The CBI says that Mr Nielson met with Mr. Bleach at Copenhagen in August, 1995, and asked him to supply 2,500 assault rifles and 1.5 million rounds of ammunition. In September that year, the men met again in Bangkok, where the order was reduced to 500 assault rifles, as well as an aircraft for executing the drop over Purulia. Mr. Nielson succeeded in organising papers which persuaded Bulgaria to sell the weapons, which its authorities thought were intended for the Bangladesh armed forces.

Though sensational, Mr. Nielsen's claims are not new — and he has offered no new evidence in support of his claims.

In a 1995 interview to the British Broadcasting Corporation, he first asserted that the Indian government had deliberately overlooked intelligence warnings of the imminent arms-drop, hoping to precipitate an insurgency in West Bengal. "An armed insurgency in West Bengal would have allowed Delhi impose direct rule on the State and oust the ruling Communists," the BBC reported him as saying.

Mr. Nielsen's claim that his escape was facilitated by the government was also made earlier by his organisation, the Ananda Marg. Bhaveshanand Avadhuta, a spokesperson for the group, said that it had "good reasons to believe Davy was escorted away by an Indian intelligence official called Dasan".

But even though little evidence has emerged to support the claims, a succession of unexplained errors by Indian authorities has meant speculation about the operation has refused to go away.

Mr. Bleach, a retired special forces officer who ran an arms firm, had notified the North Yorkshire police of the planned arms drop in August 1995. He claims to have been asked to go ahead with purchasing the aircraft and continue providing information on the plot; British authorities deny this.

Later, in November, MI5, the British intelligence service, notified India that the arms drop was imminent. A London-based British government official involved with the case said at least three meetings took place to discuss events as they unfolded, though he would not share details. The West Bengal government was not immediately notified of the threat. Intelligence Bureau personnel, it later emerged, carried out reconnaissance at possible airstrips in the area, but no on-ground surveillance was mounted by the West Bengal Police.

Mr. Nielsen himself evaded arrest at the Mumbai airport, after the aircraft was finally forced to land by an Indian Air Force jet. Investigators say he bribed authorities at the airport to escape.

A senior CBI official also said Mr. Nielsen appeared to be seeking to "muddy the waters" at the final stage of extradition proceedings. "If he has anything to say", the officer said, "we'd love to hear it — in court."

PURULIA: WEST BENGAL'S LF REGIME THE TARGET?

The revelations to an Indian television channel on Thursday by Kim Davy, a prime accused in the arms drop case over the Jhalda area in West Bengal's Purulia district on December 17-18, 1995, and by Peter Bleach who served a sentence for his role in the incident, only go to indicate that various international forces have been colluding with those within the country in targeting and destabilising the Communist Party of India (Marxist)-led Left Front government in West Bengal.

They come not just as an eye-opener at a time when the Left Front is facing its most fierce electoral battle in recent times against an alliance of the Trinamool Congress and the Congress in the ongoing Assembly polls in the State; they also serve as a reminder to the electorate waiting to exercise its franchise in the coming three phases of polling — on May 3, 7 and 10.

What Mr. Davy has had to say in his interview to Times Now TV channel is, according to the Polit Bureau of the CPI(M), "further confirmation of the fact that there was a well-planned conspiracy to use violence to destabilise the Left Front at that time... Such moves are still on."

Significantly, the arms drop occurred a year before Lok Sabha and Assembly elections were held in the State.

As the CPI(M) Polit Bureau observes "...it was a serious assault on India's sovereignty and a diabolical plot against an elected government [in West Bengal] and the Constitution."

On being asked whether the Central government knew and had authorised the arms drop over Indian territory, Mr. Davy replied: "There were political forces at the Centre who saw it as an opportune way to further their political agenda."

He continues: "You must remember that we are talking ancient history here, but in 1988 the Centre introduced Presidential Rule in Tripura [the Left Front was in power there in 1987-88] after engaging in supplying arms to different rebel groups there. The same strategy was announced publicly in the beginning of the 1990s that there was a decision to introduce Presidential Rule in West Bengal and therefore it was seen as a furthering of this agenda that arms were procured to protect local people."

Startling exposé

The disclosures to Times Now by Mr. Bleach are more startling.

Asked what could be the motive behind dropping the arms in "Indian territory in such an illegal manner," Mr. Bleach says: "To understand that you have to stop thinking of Indian territory. This wasn't dropped so much in Indian territory as in Bengali territory. The target here was the government of West Bengal. At that time, Jyoti Basu was the Chief Minister and the CPI(M) government was in power."

"The whole objective [according] to my understanding was to destabilise the government of West Bengal so that President's rule could be declared in terms of the Constitution and the State would have been ruled directly from Delhi. That could have disposed the CPI(M) government and that was the entire purpose of the job as I understand it now. I didn't understand it at that time and I have to stress that," Mr Bleach goes on to add.

As Biman Bose, CPI(M) State Secretary in West Bengal, says: "There have been moves to destabilise the Left Front government... There have been repeated demands for President's rule in the State; like the repeated efforts of the principal Opposition party in the State [Trinamool Congress] to get the Centre to impose President's rule in the State by engaging with the CPI (Maoist) in Jangal Mahal [in the south-western region of the State]."

PURULIA ARMS DROP: PRANAB RUBBISHES DEMAND FOR JUDICIAL PROBE

Senior Congress leader Pranab Mukherjee on Saturday ridiculed the demand for a judicial inquiry by West Bengal Chief Minister Buddhadeb Bhattacharjee and his party into the "political conspiracy" behind the arms drop in Purulia 16 years ago.

In an obvious reference to the Purulia arms drop case accused Kim Davy, Mr. Mukherjee, the UnionFINANCE minster, said, "An accused is levelling allegation and a hue and cry is being raised on that basis. Judical inquiry is demanded. Were they sleeping all these years?" he told reporters here before leaving for campaigning in Midnapore.

Noting that CBI had investigated the case and submitted its report, he said that the incident had taken place on the midnight of December 17-18, 1995 and after that there was a non-Congress government in Delhi for eight years.

"Indrajit Gupta of CPI, who hailed from Bengal was union home minister for two years. L K Advani was union home minister for six years. Three (assembly) elections were held in Bengal since then. This allegation was never raised," he said.

In a statement on Friday, Mr. Bhattacharjee said that a political conspiracy had been exposed in the media behind the arms drop 16 years ago and demanded a judicial probe.

CPI(M) general secretary Prakash Karat had also made a similar demand and said the inquiry should be conducted by a sitting judge of the Supreme Court.

BUDDHADEB DEMANDS JUDICIAL INQUIRY IN ARMS DROP CASE

West Bengal Chief Minister Buddhadeb Bhattacharjee on Friday demanded a judicial inquiry by the Centre into the "political conspiracy" behind the arms drop from a foreign plane in Purulia 16 years ago. "A political conspiracy has been exposed in the media behind the arms drop 16 years ago. And in this context and in the interest of internal security and the gravity of the incident therein, I demand a judicial probe," the chief minister said in a statement here. Recalling the dropping of a huge cache of arms from a plane "meant for the Anandmargis" in Purulia at midnight on December 17, 1995, the Chief Minister said, "An alert message from the Union home ministry reached the state home department a few days after the incident.

"The Central alert of December 12 was dispatched to Kolkata from Delhi by ordinary mail in a very casual manner. No advance wireless message was sent to the state on an urgent basis considering the gravity of the situation. The mail reached after the arms drop in Purulia."

Mr. Bhattacharjee, who holds home (police) portfolio, said the state government handed over responsibility of the inquiry to the CBI considering the gravity of the incident.

"All seized sophisticated arms were handed over to the CBI by the state police. All the seized arms since the days of inquiry were in the custody of the CBI," he added.

PURULIA: KARAT FOR JUDICIAL PROBE

The Communist Party of India(Marxist) on Friday demanded a judicial probe into the Purulia arms-drop case to uncover the network responsible for the incident.

"The Purulia arms-drop in December 1995 was one of the most serious instances of assault on the country's sovereignty. The investigations and the court trial proved that the arms were meant for the Ananda Marg, which was planning to use it to foment violence to destabilise the Left Front government," the CPI(M) Polit Bureau said in a statement in the wake of an interview by Niels Christian Nielsan alias Kim Davy, the absconder in the case.

The party said the judicial inquiry should be held by a sitting judge of the Supreme Court.

Addressing a press conference here, party general secretary Prakash Karat said the revelations by Davy and Peter Bleach, who was sentenced in the case, raised new questions and necessitated a fresh inquiry.

Mr. Karat said both Nielsen and Bleach made some serious charges of connivance by certain authorities in the air-drop plan and the cover up later. He said the party believed that the then Central government was "grossly negligent" in dealing with British intelligence information about the arms-drop, "but are now revising our view."

"We are not interested in scoring political points but when information was available why was the Centre not able to act and the main conspirator has still not been extradited ...the government has been tardy in getting him extradited..." the CPI(M) leader said.

The Centre sent the arms-drop information to the State government by registered post and it arrived one week after the incident, Mr. Karat said, adding that Davy gave specific details of how he had escaped from the Mumbai airport and reached Nepal.

Mr. Karat said Davy named Pappu Yadav, an MP then, as the one who assisted him and that people of Purulia saw Mr. Yadav there soon after the arms-drop.

He said Mr. Yadav, who is in jail serving a life sentence for the murder of CPI(M) MLA Ajit Sarkar, should be interrogated and the trail by which Kim Davy escaped from the country should be uncovered and the persons responsible brought to book.

The existence of the Left Front government in West Bengal has been anathema to many domestic and international forces, he said, adding that even today, the spectre of violence in the form of Maoists is being used to destabilise the Left Front.

'Killing fields'

"Even the Union Home Minister recently talked of the 'killing fields' in West Bengal and blamed the CPI(M) for them. On the other hand, Maoist violence is being aided and supported by the Trinamool Congress, which is part of the Congress-led government at the Centre."

7

R&AW and Operation Blue Star

Almost three decades after Operation Blue Star — the army operation that cleared the Golden temple complex in Amritsar of Sikh militants in 1984 — a journalist has spoken to some of the surviving dramatis personae of the event to recreate almost hour by hour what happened during those fateful six days. The documentary *Operation Blue Star — the untold story* currently being aired by Chandigarh-based television station Day and Night News run by veteran Punjab journalist Kanwar Sandhu has uncovered startling new evidence about the operation and the conduct of the militants and the security agencies since then.

Perhaps the most significant disclosures are by Balwant Singh Ramoowalia, a former Union Minister for Social Welfare, then with the Shiromani Akali Dal, who was present in the Guru Ram Das Sarai along with then Akali Dal president Harchand Singh Longowal and SGPC president Gurcharan Singh Tohra.

He relates how at around 6 p.m. on 5th June, Mr. Longowal and Mr. Tohra were coerced almost at gunpoint to declare the formation of Khalistan and how they wriggled out of it.

"Five Sikh youth with self-loading rifles (SLRs) and a metallic box that was possibly a transmitter came to us and placed their SLRs with their barrels pointing towards all of us. They told us that the 'box' is connected with Gen Zia-Ul-Haq in Pakistan. They told Jathedar Tohra and Sant Longowal to declare the formation of Khalistan, so that the Pakistani Army can launch an attack. Both Tohra and Longowal are not alive today, so I am saying this under a solemn oath of allegiance to the tenth Sikh Guru, Guru Gobind Singh, because we want to speak the truth. Sant Longowal kept completely quiet. Then Jathedar Tohra said, '*Dekho naujawano, eh jedi jang hai eh Hind-Punjab di jang hai*. This is a battle between Sant Bhinderanwale and Mrs. Indira Gandhi and that since the former is leading the battle, it will be fair to ask him to issue the statement about the creation of Khalistan.' He did not say that he will make the announcement for Khalistan. We don't know how history will judge the Akali leadership but this is the truth. The youth then left the place and never came back."

'Outrage By Army'

Mr. Ramoowalia also sheds light on an alleged execution of some 30 Sikh youth by the army — the certainty of which has always been speculated. Talking about events in the wee hours of 6th June, when the army was combing through the complex, the narrative states that a Major of the 9 Kumaon regiment lined up some 20 Sikh youth and mowed them down with a machine gun. Recalling the incident, Mr. Ramoowalia says, "The captured Sikhs appeared to be from Kashmir and didn't look like Punjabi Sikhs. An officer waved a handkerchief and they were shot dead by the Army men with bullets which were sprayed on them from left to right and then right to left. I have never seen people being killed like that, with bullets. I have been a farmer and I have cut the crop and made its bundles. The crests of these Sikh youth collapsed similarly. No one moaned or uttered anything. I know my statement will be called into questioning, but 28 years after it happened, I am going on record on this."

"The Army men were very angry, very abusive, mad with rage. Maybe they had lost their fellow Army men in the battle elsewhere in the Complex. This happened between 3 a.m. and 3.30 a.m., after the grenade blast nearby and after that it was my turn next as a part of the next group of Sikhs which was being queued up for killing. I was also told to sit down cross-legged and said my prayers. By chance, I remembered that I had in my pocket my identity card as an ex-Member of Parliament, of Lok Sabha. I flashed it and raised my hand and said, I am Ramoowalia, a former Member, Lok Sabha. I and all these persons, who are under your custody, belong to Sant Harchand Singh Longowal. We are non-violent people, [have] nothing to do with the armed struggle, we are here, just as a part of Akali Dal's peaceful morcha.' He asked me, 'what is your name?' I said, 'my name is Ramoowalia.' He asked me once again. I told him, 'I am not misguiding. Not misleading. This is my identity card. Please check it up.' God knows, the Army man was so angry, he could have just shot and killed me. But he said, 'stop'. The other Army men lowered their guns. And two to three of them came up to me... and pushed me to a side. Then the officer again asked me, 'Are you really Ramoowalia?' I said, 'I am really Ramoowalia.' He said, 'how are you here? You are not supposed to be here.' I said, 'why'? He said, 'you are supposed to be with Sant Longowal.' I said, 'Sant Longowal is sitting in the adjoining room. I have come out'."

Brigadier (retd.) Onkar Singh Goraya, who was then Col Admn in HQ 15 Corps corroborates the incident saying that Bhan Singh the then SGPC secretary also told him something similar. "He said the Army men in Darbar Sahib have done something awful. He said that some Sikh youth were lined up against a wall in the Golden Temple Complex and killed with a machine gun. He also showed me the wall in the Complex which had the bullet marks, when I went back for the second time in the afternoon," he is quoted as saying in the documentary.

Mr. Sandhu has pieced together the account with the help of interviews with some 100 eyewitnesses and officials, records from army archives, interrogation reports of captured militants and also the actual Op Instructions issued by Maj Gen Brar on the eve of the operation. Says he, "This is an honest attempt to put the events in perspective and tell the story as it happened. If in the process it upsets any one, it cannot be helped." Already the television station is getting hate mail from the Sikh diaspora in particular which is angry with its portrayal of Jarnail Singh Bhindranwale as a militant who fortified the golden temple complex with arms and ammunition.

Militants Stockpiled Weapons

In 2010 the BBC had done a one-hour documentary "1984- A Sikh story", which was never shown in India. Speaking to *The Hindu*, Mandeep Bajwa a consultant for the BBC documentary said, "This is the most authentic and credible account yet and I can see that the passage of time has emboldened many eyewitnesses to speak the truth. It exposes many fallacies like the one about arms and ammunition being planted in the temple complex by the Army. Mr. Sandhu has not only provided a rough inventory of the military hardware stockpiled inside but also detailed some instances of how they were smuggled in."

Another revelation is that Maj Gen Shahbeg Singh (retd.), the disgraced army man who joined Bhindranwale's group and organized the defences died on the evening of 5th June, before the actual battle began. He was killed by a sniper's bullet and quoting Balwinder Singh Khojkipur a close associate of Bhindranwale who survived the operation, the documentary states that he was taken to the basement of Akal Takht where he died with his head on Bhindranwale's lap. His covered body lay in a room there for a whole day until the armymen entered and cleared it the next day.

As for Bhindranwale himself —the Sikh seminary Damdami Taksal that he headed refused to accept his death for many years — he died at 8.45 am on 6th June after being shot at from an armoured vehicle as he was moving towards the 'Darshni Deodi' to offer his prayers to Guru Ram Das. His body, contrary to reports of that time, was not identified by his brother Harcharan Singh Rode then serving as the subedar major in 61 Engineers Regiment in Jalandhar, but by the police and army doctors. Says Rode, "This is totally wrong. I did not issue any contradiction because I had got to see him and paid my last respects."

Much has been written and said about the pilgrims trapped inside the complex many of who died in the crossfire. This series documents that they were actually discouraged from responding to the announcements being made by the district administration outside asking the pilgrims to come out. Apparently when five or six of them tried to come out with their hands in the air, they were shot down by militants from inside the temple complex. Their bodies lay near the ghanta ghar —where pilgrims wash their feet — on the morning of 5th June.

RAW CHIEF CONSULTED MI6 IN BUILD-UP TO OPERATION BLUESTAR

Former Research and Analysis Wing chief Girish Saxena initiated a series of meetings with the United Kingdom's Secret Intelligence Service, commonly known as MI6, in the build-up to Operation Blue Star, highly-placed intelligence sources have told *The Hindu*.

The intelligence-sharing meetings, the sources said, were authorised by Prime Minister Indira Gandhi and included at least one visit by a mid-ranking officer of the élite Special Air Service commando unit to frame an assault plan which would minimise civilian casualties.

The sources were responding to revelations that India and the U.K. cooperated on Operation Bluestar, which have led to a snowballing controversy in both countries.

Mr. Saxena, who served as chief of RAW from 1983 to 1986, and then as security adviser to Prime Minister Rajiv Gandhi and Governor of Jammu and Kashmir, now lives in retirement in New Delhi. He remained unavailable for comment despite repeated efforts by *The Hindu*.

However, another RAW chief familiar with the period, Amarjit Singh Dulat, said the consultations may have formed part of regular India-U.K. consultations on Khalistan terrorism."

"In those days," said former RAW chief Amarjit Singh Dulat, "there was very close cooperation with the U.K. on the Khalistan issue. Key leads emanated from MI6 monitoring of Khalistan leaders based in the United Kingdom." "I do not recall any specific cooperation on Bluestar, though — and I think it is probably stretching the evidence to suggest that the United Kingdom had any role to play in the actual execution of the operation itself."

Declassified revelations

Earlier this week, declassified United Kingdom government documents, first posted online by the Stop Deportations blog, stopdeportations.wordpress.com, revealed that India had consulted the United Kingdom on its decision to storm the Golden Temple in 1984.

In a February 23, 1984 letter, B.J.P. Fall, private secretary to British Foreign Secretary Geoffrey Howe, wrote that "Indian authorities recently sought British advice over a plan to remove Sikh extremists from the Golden Temple in Amritsar." "The Foreign Secretary decided to respond favourably to the Indian request and, with the Prime Minister's agreement, an SAD [SAS] officer has visited India and drawn up a plan which has been approved by Mrs. Gandhi."

"The Foreign Secretary believes that the Indian Government may put the plan into operation shortly," the letter states.

F.E.R. Butler, at Prime Minister Margaret Thatcher's office, the documents reveal, wrote to the Defence Ministry and Cabinet Office on February 6, 1983, conveying her instructions to the Foreign Secretary to proceed with the SAS adviser's request.

Limited consultations

RAW sources said the consultations were limited to seeking expert opinion on whether revanchist Sikh preacher Jarnail Singh Bhindranwale could be removed from the Golden Temple without killing pilgrims thronging the faith's most venerated institution. Bhindranwale, under pressure from a rival pro-Khalistan group, the Babbar Khalsa, had fled rooms on the periphery of the complex, taking up residence in the Akal Takht, the sanctum sanctorum.

Prime Minister Gandhi had contemplated military action against Bhindranwale since at least April, 1983, when a hit-squad he dispatched killed senior police officer A.S. Atwal. She however held back, fearing large-scale civilian killings. However, violence continued to escalate. K.P.S. Gill, Punjab's former police chief, has said that Bhindrawale's hit-squads killed 298 people between January 1, 1984 and June 3, 1984.

The SAS plan, the sources said, was rejected because Mr. Saxena was unconvinced it would work. Prime Minister Gandhi eventually handed charge of the operation to the Indian Army.

"The only plans we had were drawn up on the fly," said Brigadier Israr Khan, who led the attack under the command of Lieutenant-General Kuldip Singh Brar. "Major Jasbir Singh Raina, one of my officers, infiltrated the temple dressed as a pilgrim, scouting out hardened defences inside the temple on June 2, just one day before operations to clear the complex began."

Later, Brigadier Khan said, his troops broke open the shutters on stores near the Golden Temple, taking photographs and posters meant for pilgrims in order to develop crude operations maps.

GANDHI TOLD THATCHER ON OPERATION BLUE STAR

Former Prime Minister Indira Gandhi wrote a personal letter to her British counterpart Margaret Thatcher soon after the 1984 Operation Bluestar in an attempt to justify her decision to send army to flush out militants from the Golden Temple, the holiest Sikh shrine.

The letter, dated June 14, 1984, was made public for the first time on Tuesday as part of a British government inquiry into the role played by the UK in the lead up to the Indian army operation on Golden Temple in Amritsar to remove militants.

"It is never easy to undertake security action involving a place of worship... but this place, so sacred to the people of the Sikh faith, had been converted by terrorists into a base of operations," wrote Gandhi just days after the operation that left more than 1,000 people dead.

Also Read: UK admits 'purely advisory' role in Operation Blue Star

"We did know that arms were being collected there. But only after last week's action did we realise how vast and sophisticated these weapons were... For months a reign of terror was unleashed from the temple complex, holding all Punjab to ransom. We had no choice but to send an army unit which exercised the utmost restraint, using a minimum of force," she added.

The former Indian prime minister also shared her regrets over the fallout of the military action with Thatcher, one of her close political allies. "Many in the Sikh community have been shaken by this traumatic event. The process of healing and conciliation will take time but we shall persevere," she wrote.

The letter is among five additional documents released with the inquiry report by UK Cabinet Secretary Jeremy Heywood which would not normally have been published.

Among them is a note dated February 23, 1984, that sets out how an eight-day visit by a British military expert had helped draw out a "realistic and workable plan" to root out extremists from the Golden Temple as the Indians were fairly unprepared for action and were applying a "sledgehammer to crack a nut" principle to the whole operation.

British Prime Minister David Cameron had ordered the inquiry after documents released under the 30-year declassification rule here implied British SAS commanders had advised the Indian government as it drew up plans for a military operation on Golden Temple in February 1984. "We have taken this step because the whole investigation has been based on a commitment to the maximum possible transparency," Britain's foreign secretary William Hague told parliament in London on Tuesday as part of a statement on probe.

"We want to be as open as possible with the British public, in so far as that does not undermine the principle upheld by successive British governments of not revealing any information relating to Intelligence or Special Forces," Hague added in reference to certain names and sentences blanked out in the documents.

Describing the loss of life as part of Operation Bluestar an "utter tragedy", he revealed that the nature of the UK's assistance was "purely advisory, limited and provided to the Indian government at an early stage in their planning".

"It is also consistent with an exchange of letters between Mrs Gandhi and Mrs Thatcher on 14th and 29th June 1984 discussing the operation, which made no reference to any UK assistance," he said.

Labour's Douglas Alexander, the shadow foreign secretary, called on the government to publish the second letter as well as all other related material. Hague, however, stressed that no further documents will be made public, adding: "The investigation did not find any evidence in the files or from officials of the provision of UK military advice being linked to potential defence or helicopter sales, or to any other policy or commercial issue.

"There is no evidence that the UK, at any level, attempted to use the fact that military advice had been given on request to advance any commercial objective."

The cabinet secretary's inquiry report also points out that the fact that the UK provided some operational advice at the request of the Indian intelligence coordinator had already been put in the public domain in 2007 through a book by Bahukutumbi Raman, a former member of India's Intelligence Research and Analysis Wing.

"The UK government did send one military officer to provide military advice on Indian contingency plans for an operation at Sri Harmandir Sahib. This military advice was a one-off. It was not sustained. There was no other UK military assistance, such as training or equipment, to the Indians with Operation Blue Star," the Heywood report states.

"The UK government did not link the provision of this military advice to defence sales. The decision to help was taken in response to a request for advice from a country with which the UK had — and has — a close relationship. The military advice from the UK officer had limited impact in practice.

"The actual operation implemented by the Indian army differed significantly from the approach suggested by the UK military officer," it concludes.

THREE DECADES AFTER, OPERATION BLUE STAR LAWSUIT RESURRECTED

For 29 years, Sikhs have been fighting a case in court seeking an apology and Rs. 1,000 crore in damages from the Central government for desecrating their holiest shrine during Operation Blue Star. As the community observed the anniversary of the Army's operation at the Golden Temple with marches and prayer meetings in Amritsar on Thursday, there was some satisfaction that the three decades-old case, so far been mired in technicalities, might finally move ahead.

Unilateral decision

The Shiromani Gurdwara Parbandhak Committee (SGPC), which manages all Sikh shrines, has deposited Rs. 10 crore as court fee in the Delhi High Court where the matter has been held up since the 1980s. In March this year, the SGPC quietly submitted an affidavit withdrawing the suit, provoking widespread dismay and charges of betrayal from the community.

The decision was taken unilaterally by SGPC president Avtar Singh Makkar after the executive committee authorised him to take a final decision at a meeting. Since 1986, the case had revolved round the argument whether the SGPC had Rs.10 crore to pay the court fee. The former SGPC secretary, Manjit Singh Calcutta, who had attended many court hearings, told *The Hindu*: "In the initial years after Operation Blue Star, very few devotees used to come to the Golden Temple complex due to the presence of the Army and the police. The organisation did not have MONEY [even] to pay salaries to its employees. The then president Gurcharan Singh Tohra also felt that as a religious and charitable organisation, it was not proper for the SGPC to be paying devotees' money to the court."

On January 17 this year, the court rejected the SGPC's contention that it was an indigent organisation, and asked it to pay the fee by March 15. A day before the expiry of this deadline, the SGPC submitted the affidavit seeking withdrawal of the case. It stated the court was yet to decide maintainability of the case. Moreover, if the SGPC did pay the amount and the case was decided against it, "it will be sheer wastage of the hard-earned money donated by devotees on court fees."

But the outrage that followed this move made the SGPC deposit the money, enabling the case to move ahead. The Sikh intelligentsia feel that it is really not a question of getting monetary damages because the shrine has since been rebuilt. "For us, it has always been about getting the Central government to apologise for the operation and accept that it has hurt the sentiments of the community," said SPGC member Kiranjot Kaur. Tarlochan Singh, former Chairman of the National Commission of Minorities, said he had all along suggested that instead of seeking monetary compensation, the community should have made a symbolic fight asking for just Re 1. "That would have been acceptable to the courts and would have eliminated this avoidable delay, which gives no advantage to the community."

Mr. Calcutta Manjit Singh feels that the situation has changed perceptibly now and the matter can be resolved. "I personally think that there is a change in the mindset of the Congress because both Prime Minister Manmohan Singh and Sonia Gandhi have apologised in the last few years. It will be a statesmanlike gesture if they accept this in court too and apologise. For the Sikh community, Rs. 1000 crore in damages is not important. All we want is an acceptance from the government that a wrong was done to us."

He said that during the NDA rule, he had suggested that the BJP-led government, of which the Akali Dal is an ally, should apologise and accept that the then Congress government had erred in launching the Army operation. "But the BJP was also opposed to the idea."

CITING OPERATION BLUE STAR, PUNJAB SAYS CENTRAL FORCES INSENSITIVE

Punjab has rejected the Union government's plan for the National Counter Terrorism Centre, calling it a direct interference in India's federal structure. The NCTC will allow the Centre to tactically intervene in internal matters of the States, Deputy Chief Minister Sukhbir Singh Badal said on Wednesday.

Speaking at the annual conference of Chief Ministers on internal security in Delhi, he said the Centre reduced the session to a ritual, where not a single proposal or suggestion made by them had materialised in the past four years.

Mr. Badal said the NCTC plan appeared an attempt at raising the bogey of internal security with the States. For, there was no NCTC when Punjab fought back decade-long terrorism on its own. If allowed to operate under the proposed NCTC, Central forces would descend on the States without assessing the sensitivity of the situation, launch an operation and move out, leaving them to bear the brunt.

'Insensitivity'

Citing Operation Blue Star, which Central forces had launched on the martyrdom day of Guru Arjan Dev, leading to the death of a number of innocent pilgrims trapped inside the Golden Temple in Amritsar, he said such gross insensitivity on their part, instead of tackling the situation, complicated it.

So, Mr. Badal said, the Centre should provide full backing to the States in their plans for modernisation and capacity-building. Rather than holding the annual meeting without any substantial result, the Union Home Minister should convene State-specific meetings individually and tackle particular problems.

Demanding a security package for Punjab, Mr. Badal said the Centre must support the border State in its battle against cross-border terrorism, smuggling of drugs, counterfeit CURRENCY and Web-based militant propaganda. Though "we have been able to eliminate militancy, efforts are made by forces across the border to foment trouble in the State."

A record seizure of drugs by the Punjab police in the last few years underlined the fact that the Border Security Force, working directly under the Union Home Ministry, had failed to seal the border effectively. Punjab ended up paying the social cost of increased drug flow, with the Centre remaining a mute spectator, he said.

BLUE STAR ANNIVERSARY LARGELY PEACEFUL

Except for a minor skirmish between the Shiromani Gurdwara Parbandhak Committee (SGPC) task force volunteers and the activists owing allegiance to the pro-hardline leader, Simranjit Singh Mann, the 29th "ghallughara divas" (genocide day) on the anniversary of "Operation Bluestar" passed off peacefully.

While people participated in special prayer services at most gurdwaras across the region, the main function was organised at the Sikh community's apex religio-temporal seat, the Akal Takhat, within the Golden Temple complex. The Jathedar, Giani Gurbachan Singh, who presided over the function bestowed 'siropas' (robes of religious honor) on the members of the families of militant leaders including Sant Jarnail Singh Bhindranwale, former president of the All Indian Sikh Students' Federation, Bhai Amrik Singh and General Shahbeg Singh, who have been declared as the "quami shaheed" (martyrs of the Sikh nation). The three were killed in the army action.

The oppressive heat notwithstanding, as the Golden Temple complex witnessed an extra rush devotees since morning, while the Amritsar district administration as well as the SGPC had beefed up the security arrangements in the walled city and the shrine, respectively. As per a previous decision, no event was organised at the recently constructed Bluestar memorial, whose basement remained locked, while recitation of the holy scriptures continued on the ground floor. The new building constructed near the Akal Takhat by the religious seminary, Damdami Taksal, evinced interest from the devotees. The building has been constructed at the spot where Sant Bhindranwale's bullet riddled body was recovered after the completion of the army action. He was the then chief of the Taksal.

In his address from the Akal Takhat rampart, the Jathedar appealed for Panthic unity to ensure justice for the Sikh community, which had been smarting under the impact of the wounds caused by the attack on the holiest shrine as well as massacre of the Sikhs in Delhi and other places following the assassination of the then Prime Minister, Indira Gandhi. He said that it was ironic that while those who were found guilty for assassinating Mrs Gandhi, had been executed long ago, the cases of justice for the Sikhs were still hanging fire.

Giani Gurbachan Singh said that while Devinderpal Singh Bhullar, who was convicted on the basis of a "confessional statement" extracted under duress by the police, continues to be on the death row, Congress leaders, like Sajjan Kumar and Jagdish Tytler, were acquitted despite major evidence against them for their involvement in the November 1984 anti-Sikh carnage. The Jathedar appealed to the community that without forgetting the events of 1984, the believers must remain vigilant against recurrence of such incidents that hurt the collective psyche.

On the other hand, for a brief period it was a free for all as swords were bared, fists flew and turbans knocked off, when the volunteers of the SGPC task force and supporters of former MP, Simranjit Singh Mann clashed near the entrance of the Akal Takhat. The trouble ensued after Mr Mann's supporters who were raising slogans in favour of "Khalistan" were stopped by the SGPC task force on the pretext that the hall inside that Takhat building was packed to it capacity.

Mr Mann's supporters were also seen heckling the SGPC president, Avtar Singh Makkar, who chose to use the far end entrance to pay obeisance at the Akal Takhat and leave in quick succession. He did not stay long during the ceremony. Senior leaders of the ruling Shiromani Akali Dal, which controls the SGPC, were also conspicuous by their absence.

Later, talking to reporters Mr Mann said that the struggle for a separate homeland for the Sikhs was a long drawn one and he would continue with his efforts. He also announced his plan to visit China and impress upon the government there to recognise the demands of the Sikhs, who needed support from a power, which had more muscle and resources than the Indian State. He said due to internal compulsions, Pakistan was no longer in a position to sustain its support to the Sikh cause.

8

Role of Indian Intelligence Agency in Secession of East Pakistan

BANGLADESH NEW YET OLD STORY

Nothing new in the fact that the Indian central intelligence agency, R&AW had plan for secession of East Pakistan from the day one of 1947 but also that Bangladesh came into being as the essential by product of the design. A US NGO report has only now in August 2009, too late if not too little, made the fact public in their document. Those who are already aware of the R&AWs machination in early 1950s and 1960s as there are in Jyoto Senguptas memoirs (A History of Freedom Movement of Bangladesh 1947-73: Some Involvement, 1973) and Ashok Rainas documentary evidences (Inside RAW) had no surprise in the matter that David Miller Smith made on the 24th August 2009 a three- page item I have received from the internet.

Bengal Muslims

If one would honestly recall the background history and movement of the Bengal Muslims during the colonial British period it would really appear to be a matter of surprise that the people who popularly supported the Muslim League since almost about the inception in 1906 and overwhelmingly voted for one Pakistan in the 1946 general election could have gone all the way out for secession from Pakistan and seriously sought for independent

Bangladesh just after 24 years in 1971. Five years back in September 1965 when Pakistan-India war broke out, and I was then posted at Rangpur district town about 200 miles north west of Dhaka, teaching in a college there, I recall thousands in that small district town thronged into the streets to cheer up the EPR (East Pakistan Rifles) and the Pakistan Army units posted in the locality to crush India at the war. Did I see anything wrong of the peoples sentiment against Indian aggression or were the people wrong about their sentiment for Pakistans unity and integrity? The 1971 26th March army Operation Searchlight changed no doubt many calculations, but was that deep enough to bring secession and cut Pakistan to size for ensuring all strategic benefits to the regional Indian power? Was there no scope to look for dignified and respectful way out except secession?

Why not all Bengalis?

Some other relevant questions could also be raised. Indians had been eager to get Bengalis of East Pakistan in 1971 freedom from Pakistan; very lofty proposition indeed, and a pious wish of them. Why could not they have had even before 1971 the same pious wish for freedom of the Kashmiris who have been brutally enchained against their free will by Delhis occupation armed force for over six decades now figuring about half a million and since 1947? How about the independence struggle of the people of the so known â€˜seven sisters of the eastern India who remained close to and neighbor of former East Pakistan and now Bangladesh? Why cant Delhi let the 80 million Bengalis of West Bengal and Tripura living adjacent to Bangladesh territory could go off its control and suzerainty to form together still greater Bengali State of about 230 million people with Bangladesh after 1971? I am sure, Delhi had no easy, much less satisfactory answer to all the questions above, and so they cannot rationalize secession of East Pakistan in 1971. That means the R&AW and Delhi had other rotten rats in the bag.

RAW in FASs findings

Possibly the Federation of American Scientists (FAS), established in 1945 and having had its wide- ranging members of high credentials including members from among 45 Nobel Laureate cannot be given a dam to the RAW involvement in the secession of East Pakistan since 1947.

That is however not to suggest that thousands of sincere and dedicated Bengali freedom fighters had no mean contribution in snatching independence of Bangladesh in 1971. Even so, say for example, foreign dignitaries like US Senator Ted Kennedy, etc. who among the rare Americans supported the Bangladesh movement in 1971 against his country's policy not to support secession need be seen in still wider context.

Exuberant and yet immoral Kennedy

By the way, Kennedy coming of a political family had been an exuberant Democratic Party youth of 39 years in 1971. He was well known to be licentious as many of the Democrats used to be. He was warned several times by police for fast driving quite possibly under alcohol. In moral turpitude he was a close friend of such elements. Though free sex and alcoholism are no public crime so long those remained indoors, but other moral turpitude could have been with these vices. Once in 1969 he had a young girl with him at a nightclub, possibly drank heavily, and then driving himself to their destination. Unfortunately, the car met an accident he was driving when the car overturned and fell in a water pool by the road from a small bridge on the road. He swam across from inside the car, went away caring nothing for the girl (Possibly call girl), not even informed the police to rescue her from the drowned car. That incident should speak well about the standard of his morality and ethics. Exuberance for a subject matter and feeling for a great humanitarian cause in association with high morality and ethics are different matters.

Akhanda Bharat

Coming back to the main theme and concentrating in deeper aspect of the subject, irrespective of what other parties contemplated, India went absolutely ahead with her own design that she set right in 1947 just as Pandit Nehru made clear in May 1947 in a letter addressed to one of the Congress leader Ashrafuddin Chowdhury of Comilla. He stated clearly in the letter addressed to him dated 23 May that they had accepted the condition of partition of India (and for founding Pakistan) for the time being as a strategy for â€˜reuniting once again India sooner than latter. If one would recall further the first spontaneous response of the Indian Prime Minister Indira Gandhi on the 16th December 1971 announcing the defeat of Pakistan Army to the Indian victorious Army General Arora, ˜HAME HAJAAR SHALO KA BADLA LE LIA (we have taken the historic revenge of one thousand years)!

Revenge of Delhi against Pakistan alone? Revenge for whose freedom? Bengalis freedom? Why not for the Bengalis of West Bengal and Tripura along with Bengalis of Bangladesh?

Weaker Pakistan

It is true that for the R&AWs success Pakistan has been weakened after 1971 that emboldened India more than ever before 1971. Even so, there is some competition for rivalry at least in possession of nuclear arms by both Pakistan and India. How does Bangladesh fare in this matter? Not even a nuclear power station! Unequal and subservient That Bangladesh does not fare equal with India is a fact of reality. In addition, Delhi had imposed whatever they wished against Dhaka ranging from the Constitution tailored to their need and goal for AKHANDA BHARAT or reunited India to complementary and supplementary economy, to subservient foreign relations, to educational and cultural policies, to internal security and even in matters of national secrets, if there is any. The 25 year treaty Bangladesh made in Dhaka on the 19th March was clearly a treaty of subservience, particularly by the power of the clauses 8, 9 and 10 (though it lapsed in 1997, the hangover remains in other areas like the 30 year treaty for water sharing and the CHT treaty of gross inequity among citizens of the same country and locality). The vulnerability of maritime boundary with the neighbours, inefficient border protection, control of water flows at the upstream of the 54 rivers by India at their free will and causing all disadvantages to Bangladesh remained as Achilles Heel of Bangladesh making independence a play toy in the hand of Delhi so much so that more of the capable ministers like AMA Muhith made on the 18th August a frank but guarded remark, The sovereignty of the nation is limited now due to different reasons (The weekly Holiday, 29 August 2009). Unfortunately he did not elaborate or had to guard his job by not elaborating and indict the R&AW and Delhi.

15th August 1975 gains now lost

Some patriots, particularly in the Army, who having had discovered the mockery made a decisive blow and ousted the puppet government on The 15th August 1975, because the game of secession had no clear sanction of the people but engineered and well managed by the R&AW.

Unfortunately, the gains accrued were almost all lost and brought the R&AW once again in to the driving position so much so that many sensible men and women consider the High Commissioner of Delhi now in Dhaka in full control of internal affairs of Bangladesh since January 2009.

Affluent few and millions have not

Well, in the kind of independence a microscopic minority have thrived but the main slogan for the masses in millions for emancipation remain a matter of illusion in the last four decades so much so that nearly half of the 150 million live under the extreme poverty line. And if the domination of Delhi and the policy of Akhanda Bharat of the RAM RAJ or caste ridden division between man and man continues to be operative, the R&AW would continue to keep its hold for such perpetual discrimination, and so would continue to remain illusive the real freedom for all in Bangladesh.

ZIONIST-HINDUTVADI PLOTS AGAINST PAKISTAN AND INDIAN OCCUPIED KASHMIR

How much more blood must be shed in the Islamic Republic of Pakistan before it is understood that the country is being torn to shreds from the inside out in a foreign-sponsored 4th Generation Warfare (4GW) campaign? How many more men, women and children must have their lives snuffed out before a little, not a goliath amount, but a little bit of attention is given to this horrific and blood-soaked crisis that deepens every day? The geopolitical implications will surely be drastically and negatively groundbreaking if Pakistan plunges into the unadulterated chaos that its enemies, "Israel," Hindutvadi India and the Zionist-occupied United States government, so passionately and stridently desire.

Pakistan, the "child of Jinnah, child of 'Allama Iqbal", indeed stands on the brink.

Hideous sectarian killings across the country since the start of April have revealed the deepening of a plot engineered by the forces of Zionism and Hindutva, its ugly first cousin. Whilst the Zionist media coverage has typically and predictably attempted to depict the bloodshed as "Sunni-Shi'a" strife, it is quite revealing that its coverage has been scarce, with the weight of reportage being carried by Pakistani and Iranian outlets. Additionally, the evidence will show that this sectarianism, drenched in innocent blood, is a product of the Zionist-Hindutvadi nexus, with the sect of oppressed being butchered by the sect of the supremacist oppressor.

It began on April 9th, 2012 in Pakistan's turbulent Balochistan province, in which masked gunmen on motorbikes carried out a monstrous shooting spree, leaving 14 dead, including 2 women, and dozens of others wounded. The worst of the attacks was in the Hazara region at the busy Prince Road of Balochistan's largest city and provincial capital, Quetta, where four 4 Shi'a Muslims were shot dead, and then, according to later reports in theUrdu and Farsi presses, brutally mutilated, thus betraying the entire bout of killing as a specific targeted operation. As the casualties count rose to 6 dead and 3 critically wounded the following day, this notion was further confirmed when Balochistan Chief Minister Nawab Aslam Raisani fired six police officers for complicity in the Quetta bloodshed, revealing that the gunmen had "men on the inside" to protect their operation from scrutiny.

In response to the martyrdom of 6 members from their community, leaders of the Hazara Democratic Party, which represents the interests of the downtrodden Shi'a of Balochistan, adamantly and eloquently blasted the cowards that perpetrated the crimes. They then called for a 10 day strike, from April 20th to April 30th, demanding justice for their fallen. Hundreds across Balochistan honored the call by protesting and shutting down shops, markets and educational institutions. Following the call of the Hazara Shi'a, more than 2,500 Pakistanis, Sunni and Shi'a alike, in a beautiful and righteously ferocious display of solidarity, hit the streets of Islamabad right outside the Pakistani Parliament building and chanted for the government to take action against the persons responsible for the violence. Similar demonstrations were held in Quetta, Muzaffarabad and Multan, where protesters supplicated for the Quetta victims as well as the more than 2,000 Shi'a murdered in the Kurram region of the Federally Administered Tribal Areas (FATA) since 2007.

The ostensible drift from blood-spilling to reconciliation and justice came to a screeching halt on April 14th, 2012 as Balochistan was rocked by tragedy yet again. In the same fashion as the previous killings, masked gunmen laid siege to the Barori Road, Sabzol Road and Qali Chaltan Road areas of Quetta, firing shots at dozens of Shi'a Muslims, wounding at least six and murdering at least 10. This death toll would rise to more than 30, including women and children, and on April 16th, another Shi'a was gunned down by masked men in Quetta, prompting Chief Minister Raisani to thunderously call for assistance from the Pakistani Army and declare that if something wasn't done to stop the atrocities soon, Balochistan would descend into civil war. The obvious purpose of the cowardly masked men's assault was to foment division but this has failed as Sunni-Shi'a unity protests raged forward, with women and children participating, and not just in Pakistan, but all over the globe, including America, the UK and Canada.

On April 18th, 2012, Imran Zaidi, the vice principal of Jinnah Polytechnic College and a Shi'a Muslim himself, was shot dead by masked men on motorbike in the important Pakistani port city of Karachi, once again betraying the signature of an intelligence agency's "targeted" operation. The massacring of Shi'a continued three days later, with another attack in Quetta by masked gunmen on motorbike, shooting two more innocents dead. Immediately following the latest round of killing in Quetta, protests across

Pakistan erupted yet again with Shi'a marching through the streets of Lahore and joint Sunni-Shi'a rallies being staged in Abbottabad, Quetta, Rawalpindi and other cities. A sigh of relief swept through the nation when Pakistani security forces stormed the outskirts of Quetta in a search and seizure operation, uncovering the safehouse where the carnage has been launched from. Three terrorists were detained, all of whom were connected to none other than the Tehrik-I-Taliban of Pakistan {TTP}, the infamous and brutal proxy force of Pakistan's enemies. Two more fighters connected to the Shi'a killings, Ali Sher Haidri, spokesperson of the TTP-allied Lashkar-e-Jhangvi (LeT), and his well-known terrorist accomplice, Gul Jan aka Hafiz Naseer, were killed in another operation.

Created in 2004 to coincide with the inauguration of the CIA's genocidal drone program, the TTP is the perfected product of an ongoing, incestuous relationship between the Zionist entity's Mossad, Hindutvadi India's RAW and the CIA itself. One month after Mossad's false flag attack on September 11th, 2001 and just three days after the criminal invasion of Afghanistan, two "Israeli" agents with Mossad and IOF connections were apprehended inside the Mexican Congress building of Mexico City with 9mm handguns, nine grenades, C-4 explosives, three detonators, 58 bullets and fake Pakistani passports. Their mission was to destroy the Mexican Congress. This false flag operation was intended to be phase II of 9/11, triggering a NATO invasion of Pakistan. When the mission failed, the Zionist entity activated its contingency plan: the Dragon Policy, an updated version of the Oded-Yinon-strategy of balkanizing Islamic states on ethno-sectarian lines.

Following the failure in Mexico City, the upper echelons of Zionist and Hindutvadi intelligence erected four new clandestine agencies solely dedicated to the destruction of Pakistan, which would immediately be followed by the originally-intended aggression to denuclearize the bulwark of Islamic Resistance to international Zionism. The highest ranking figures of Pakistan's financial, political, military and religious sectors were on the hitlist and every successful attack was meant to soften up Pakistan for the imminent invasion. Using funds from the unlimited bank accounts of Mossad and Aman, the Zionism-Hindutva intelligence nexus established 57 training camps across the northern frontiers of India and occupied Kashmir, kidnaping and blackmailing Resistance fighters to fight against their own comrades while simultaneously forging alliances with mafia bosses and narcotics tycoons and training them in the finer arts of black ops. These various elements served as the perfect puppets as any rebellion whatsoever would be punished with imprisonment in the harshest of dungeons. These men would be deployed into the field under Mossad-RAW guidance and once an attack was completed, the Zionist media would report it as an 'act of Islamic terrorism.' If a 'checkout' was not secured, senior operatives of the "Israeli"-Indian nexus would finish the job.

The CIA and its corporate offshoot, Blackwater (also known as Xe and Academi), a festering bubble of anti-Arab, anti-Muslim hatred and deep pro-"Israel" sentiments, joined the Dragon Policy in 2004. Blackwater agents became integrated with the ordnance units of Mossad and RAW. The CIA took a more proactive role by absorbing past recruits of the Zionist-Hindutvadi intelligence training camps and cultivated them into a powerful proxy force that specialized in guerilla tactics and sectarian warfare: the Tehrik-I-Taliban. The name was chosen for a very sinister, deceptive and specific reason: to despicably smear the real Taliban in occupied Afghanistan, a Resistance force fighting a '7arb el 7uriyeh (war of liberation)' against NATO invaders. It is classic psychological warfare. The Taliban, not surprisingly, has rightfully maintained its innocence and condemned Blackwater for being behind bombings and other provocations in Pakistan. Since its birth, the Tehrik-I-Taliban has murdered or wounded more than 100,000 Pakistanis and caused more than $50 billion in damage to Pakistan's economy. Prior to the most recent operation, the Pakistani army had already arrested hundreds of TTP militiamen. Proxy agents detained in Waziristan, Karachi and Balochistan were caught with sophisticated, military-grade weaponry manufactured by the Zionist entity.

This string of horrible bloodshed that claimed the lives of dozens of Shi'a Muslim men, women and children, along with several other Sunni Muslims who stood in solidarity with them, is reminiscent of a very similar Mosssad-RAW-backed TTP campaign that targeted prominent Shi'a and Sunni figures back at the start of 2012. Tearing Pakistan apart through the"sectarianism" of the Zionist-Hindutvadi Dragon Policy, leaving it open to denuclearization and balkanization and eventually fomenting the ethnic cleansing of all Muslims from occupied Kashmir is indeed the "endgame" plot.

The innocent Pakistani blood spilled throughout April means nothing to the supremacist oppressors or their quislings, and this is why the great Pakistani luminary Syed Zaid Zaman Hamid declared recently in one of his highly-regarded email bulletins that, "Pakistan's entire political leadership is now compromised to the enemies. These ones want to compromise our honor, faith, history, land and water," in fact, he continues, "the idiots we have today have already auctioned it." And if the Pakistani Armed Forces didn't perform the way that they did in Quetta, ending the catastrophe (at least for now), it is safe to say that the enemies would be further along in their plot.

OCCUPIED KASHMIR: THE "ISRAELI"-INDIAN "TRAINING GROUND" FOR ANTI-PAKISTAN WARFARE

It is a well-established and internationally-accepted fact that the Zionist entity's military and arms industries are imposing behemoths and completely unrivaled on the international stage. What is lesser-known, or rather, ruthlessly suppressed by the Zionist media as a means of keeping it far from the global public's vision, is that "Israel" uses the occupied West Bank (and by extension, occupied al-Quds) as a laboratory for the research and design of its newest weapons. Zionist occupation forces use the hundreds of Palestinian villages in the West Bank as 'live' training grounds. Though this certainly qualifies as "sick," as well as "twisted," as Zionism always does, it gets even more depraved.

On March 27th, 2012, the undercover unit "Israeli" military intelligence known as Duvdevan, which is responsible for the murder of beloved Palestinian-Jewish activist and director Juliano Mer-Khamis, as well as a laundry list of extrajudicial executions dating back to the First Intifada, savagely murdered 28-year old Palestinian Rashad Shawakha and badly wounded two of his brothers during a 'live' training exercise in the occupied West Bank Village of Rammoun. For the very first time, as revealed herein by Mask of Zion's ever-vigilant and exceptionally reliable Kashmiri sources, it is now known that these slaughterous and utterly insidious "live training" and "weapons testing" methods have been exported to occupied Kashmir, the land known as "heaven on Earth" illegally occupied by Hindutvadi India for 65 years and already infested with Zionist-Hindutvadi training facilities for the Dragon Policy.

For the historical record, it is integral to note "Israel" is not only the top defense supplier for India, but the usurping Jewish entity and the Hindu supremacist regime are deeply intertwined at every level, from military to intelligence, economics to politics, communications to ideology, and this cooperation not only amounts to a frightening hegemonic agenda, but tens of billions of dollars in profit generation. The two occupation entities even collaborate in ventures to provide cell phone gaming services for their brainwashed populations ; the Zionism-Hindutva relationship is one of the utmost geopolitical intimacy.

Ironically and ominously, although perhaps not unintentionally, one year ago to the day before the barbaric massacres of Shi'a Muslims began in Balochistan, a terrible tragedy occurred in occupied Kashmir that shocked and saddened the people of the Vale. Maulana Showkat Ahmad Shah, a pseudo-controversial but much-loved cleric who served as the President of Jamiat-e-Ahl-e-Hadith, was murdered in a brutal bombing during Friday prayers at a mosque in the Kashmiri capital of Srinagar. So abominable was the attack that it prompted a harsh condemnation from every major Azadi (Kashmiri word for freedom) organization throughout occupied Kashmir, as well as the government of Pakistan.

Survived by his wife, son, daughter and mother, well-known across the Vale for his honesty and dedicated to the cause of Azadi by all accounts, Shah had been arrested and imprisoned under the Hindutvadi occupation's draconian Public Safety Act (PSA) many times before, and on at least two other occasions, there were attempts on his life. Dating back to 1990, a phenomenon known in occupied Kashmir as the "unidentified killings" has claimed the lives of dozens of doctors, academics, clerics, philanthropists, Resistance fighters and human rights activists, including Muslims, Pandits and secularists. All evidence indicates that RAW's "Special Operations Group" is behind the murders. The most recent target, prominent religious scholar Peer Jalaludin, barely escaped with his life after being critically injured in a shooting. Azadi leaders pegged Hindutva as the culprit behind the barbarous attack, stating that it was trying to divide the Kashmiri struggle through sectarianism.

Mask of Zion contacted several Resistance fighters and Azadi activists to learn more about these disturbing developments and what was uncovered during these inquiries turned out to be nothing short of shocking and enlightening, as well as exceedingly revealing. One 46-year old veteran fighter, who, for security reasons asked to be identified by the alias "Rehman Zahid", an eloquent man with vast experience in the Azadi struggle as he fought the Hindutvadi occupation head-on during the epic Kashmiri intifada of 1989, explained to Mask of Zion the nature of what he referred to as "SOG," the RAW Special Operations Group, "The SOG is active all over Asia. It is in Pakistan and Sri Lanka, Nepal and Indonesia, Malaysia and as far north as Japan. It is responsible for assassinations and surveillance and anything else the accursed Hindutvadis ask for, may ALLAH (SWT) curse them to the hellfire. Curse them! But here, here in Kashmir, my beloved that they have been raping since before Palestine was raped by the Zionists, my land of treasures and rushing rivers, the SOG is subordinate to an 'Israeli' named Katzir. You know him, yes?"

"Yes, of course," I stated anxiously. Brother Rehman continued, "The 'Israeli' is governing dozens of SOG agents with full approval from the Hindutvadis in Delhi. He's integrated them into the units made up from the elites of his counter-terrorism team and calls these units the 'Yehoshua Brigades', named after Joshua from the Jewish Bible who was a destroyer and murderer of children, a rapist of land and thief of Canaan." Understandably, brother

Rehman paused to take a deep breath. I saw this as an opportunity to interject, "These YehoshuaBrigades aren't regular military units then, and the operations of the SOG aren't normal operations, are they?"

The Kashmiri Resistance veteran paused for yet another moment and solemnly said, "No, not at all. He named them 'Yehoshua' because these units are those which will carry out Operation Kurukshetra, a creation of Katzir and the monster, {a reference to former Indian Chief of Staff, General Sundararajan Padmanabhan, the Hindutvadi fundamentalist architect of the Cold Start Doctrine known throughout the Vale for his sadistic brutality during the 1989intifada and the immediate years after it } to cleanse us from the land like Joshua and his band of murderous yahoud cleansed Canaan thousands of years ago and cleansed it again in 1948. Every assassination that they've done in the last decade was designed to provoke us into attacking them so they can make Kurukshetra a reality. Curse them, ya ALLAH (SWT)! Curse them!"

"Would we be providing a most accurate description," I asked, "if we designated the YehoshuaBrigades, in all honesty, as an exclusivist Jewish-Hindu death squad motivated by the Zionist-Hindutvadi leadership's desire to attain ethno-religious and regional supremacy?"

Without hesitation, Rehman Zahid, a man who had stared into the eyes of the cruel and torturous Indian soldiers who waged an unrelenting campaign of brutality under General Sundararajan Padmanabhan during the 1989 Kashmiri intifada, stated, "Absolutely. Absolutely accurate. Supremacist is what they are. Exclusivist! Separatist is what they are! They call us separatists but they are the separatists! They want to separate from all other humans, with yahoud being with yahoud and Hindutvadi being with Hindutvadi while we suffer under them. They think of us as dogs. Enemies of Islam! Enemies of humanity! Enemies of the Kashmiri, the Palestinian, the Lebanese! Enemies of Pak Sarzameen and the dunya and even the Akhira! Curse them! May ALLAH (SWT) give us strength to fight them and free ourselves from them!"

Brother Rehman had become thoroughly flustered and quite upset and therefore, out of respect for and empathy with him, I had no interest in prolonging the discussion for any serious length of time; I only had one more question, "Akhi Rehman, this information is indeed critical and it most certainly explains a great deal, but what about the assassination of Maulana Showkat Ahmad Shah? Was the attack on his person a hyper-specific one?" Obligingly, Rehman replied, "Not necessarily, but, Shahid Maulana was their highest-level target yet methinks. Every killing that Yehoshua does is based on its objective of driving us into civil conflict and hurting our struggle. There is another goal for them though. The explosive used to martyr Maulana, a sticky bomb, was a modified, military-grade version of what was used in the July 11th Bombings and the Malegaon Blasts {which were false flag attacks carried out by the Zionist entity's Mossad and Hindu supremacist elements close to RAW to set the stage for their 'masterpiece' false flag on 26/11, an event that 'changed the game' in the region }. Yehoshua and Katzir's regular units and even the Hindutvadi occupiers themselves are always testing new weapons for 'Israel,' may ALLAH (SWT) curse them to the hellfire. We're like their playthings."

He continued, "We are armed with minimal weaponry here, anyone can see this, even the blind ones. What do they need heavy automatics, laser snipers, drones, pistols with silencers and upgraded military explosives for? Are they that afraid of our youths who pelt stones? Do our stonepelters make the Hindutvadi and the Zionist shake? Maybe we do, by ALLAH (SWT) maybe we do, but this isn't the reason. They're testing their weapons here because these are the weapons that they use on Pakistan to make chaos there. Kurukshetra is the final part of the Hindutvadi Cold Start Doctrine. Before Yehoshua Brigades are released to make the Kashmiri Nakba, India and the NATO will have turned Pakistan into Iraq or Libya with Cold Start, may ALLAH (SWT) damn them for what they did to those countries and the orphans.Yehoshua Brigades try and kill Peer Jalaludin, may ALLAH (SWT) protect him, right outside of his home with new never-before-seen heavy automatics and not even a week after this, not even a week by ALLAH (SWT)! These weapons are in the hands of TTP terrorists who spill blood in Pakistan. This is the pattern. Like I said, we are their playthings. Pak Sarzameen is the grand prize in all of this. We are the small fish."

We thanked brother Rehman for his time and Muhammad Isma'il Bhat, the friend of mine who put me in contact with the Kashmiri veteran, told me in the clearest language possible that everything he said was on the money. I relayed this information to human rights activists "Zubair Zubair" in Srinagar and "Abu Ghulam" in Kupwara, who had contributed to a previous earth-shattering Mask of Zion report on Pakistan and occupied Kashmir, and both of them reaffirmed it unequivocally. It was official. Occupied Kashmir, "heaven on earth," was being used by the Zionist entity and Hindutvadi India as a "live testing ground" for Dragon Policy directives that will later be carried out in Pakistan, with every killing aimed at disuniting the anti-occupation movement and delivering it into the jaws of destruction. The Kashmiri people were at the mercy of a Jewish-Hindu death squad known as the Yehoshua Brigades, with every moment of their day passing by as if it were their last.

Once again, this shows that the fates of the Pakistani and Kashmiri people are interlocked andAzadi for one from the machinations of Zionism and Hindutva means Azadi for the other. And with Kashmiris preparing to launch another intifada, like they always do, seemingly every two years, there is no doubt that this theme will permeate protests and marches. There is no doubt that they will demonstrate for dignity and with vigor. There is no doubt that the oppressed people of the Vale will not relent until their Zionist and Hindutvadi tormentors are brought to justice and it is known that Kashmiris are the "playthings" of nobody.

THE SEPARATION OF BALOCHISTAN AND OPERATION BLUE TULSI EXPOSED

Aforementioned Pakistani luminary Syed Zaid Zaman Hamid has repeatedly likened what the Islamic Republic of Pakistan is fighting today to the "Agartala Conspiracy" of 1971, in which RAW, CIA and elements of the Soviet KGB created, trained, armed and funded the Mukti Bahini terror squads that broke down the

Pakistani Armed Forces through guerilla warfare, setting the stage for the breakaway of East Pakistan, known today as the nation-state of Bangladesh. There was no "popular Bengali uprising" as it often parroted by Zionist media and academia, there was a plot of subversion directed by the enemies of Pakistan with Hindutva at the head. This is further confirmed by the Syed in which he points out that the Pakistani Army was 10,000 miles away from base, fighting 4GW in the streets while India invaded from three sides with 500,000 soldiers and 250,000 more Mukti Bahini terrorists. The "soldiers of Jinnah, soldiers of 'Allama Iqbal" were overwhelmed, and the dismemberment of Pakistan was completed. It is highly critical to add that the Zionist regime assisted India in this plot, as balkanizing and weakening Islamic states is a cornerstone of the Jewish supremacist entity's foreign policy, as discussed earlier in the chronicling of the TTP's history.

The modern day "Agartala Conspiracy" can be accurately referred to as the "Balochistan Conspiracy." Just before the Mossad-and-RAW-backed TTP began carrying out its latest massacre of Shi'a Muslims in an attempt to drum up more sectarian tension across Pakistan, which, by the way, is an ongoing machination as Pakistani security forces just stopped another possible attack on Kirani Road in the Hazara Town neighborhood of Quetta, a tragedy of literally mountainous proportions struck the Pakistan nation on April 7th, 2012 as an avalanche on the famed Siachen Glacier, a massively geopolitically important strategic nexus and integral, ancestral part of Pakistan's national and cultural heritage, slammed down upon the headquarters of the 6NLI battalion of the Pakistani Army, trapping at least 139 soldiers and civilians underneath the frozen tundra. Illegally occupied by Hindutva in April 1984 in Operation Meghdoot, Siachen is once again back in the spotlight and reminding Pakistanis that the agreement signed between their nation and India in 1989 to secure the withdrawal of all forces has been consistently ignored because of Hindutva's intransigence and its dreams of supremacist expansionism.

Developed by the Indian Armed Forces' Western Command in 1978 and actually engaged by the Indian Armed Forces' Northern Command, Operation Meghdoot was launched to capture

Karakoram Highway (KKH), therefore disconnecting Pakistan's road links to China, as well as to capture the passes that connect Siachen Glacier to Gilgit-Baltistan, therefore confiscating yet another piece of land historically linked with occupied Kashmir. This act of Hindutvadi criminality was meant to bog down large numbers of Pakistan's best and brightest forces in high altitude warfare, while continuing the build-up towards the initiation of the Cold Start Doctrine and Operation Kurukshetra. Moreover, what is additionally and most importantly attached to the "bog-down" strategy of the illegal Indian occupation of Siachen is its policy of supporting terrorists in Balochistan, specifically the Balochistan Liberation Army (BLA) and of course, the TTP. Hindutva and its allies in the CIA and Britain's MI6 are also funding, arming and training the Baloch Liberation Front (BLF), Baloch Republican Army (BRA) and Lashkar-e-Balochistan. RAW and CIA have set up 60 training camps across Balochistan to maintain this destabilization, which is occurring just 100 miles east of Iran.

Where there is Hindutva, Zionism is never too far away and this is especially true for Balochistan. Another vital asset of the gaggle of secessionist-terrorist elements being utilized by the enemies of Pakistan to plunge the Islamic nation into devastation is the shadowy group known as Jundallah, which has been trained and armed by Mossad for years now. Mossad has used Jundallah against the Islamic Republic of Iran on numerous occasions in bloody, bloody slaughterous attacks and even Ayatollah Sayyed Ali Khamenei himself has made mention of it.

In its typical use of its classic and favorite tactic, the false flag, operatives of Mossad have also posed as CIA agents and flaunted wads of American cash and US passports to recruit Jundallah into carrying out bombings against Iranian women and children to terrorize the Zionist entity's most hated enemy. While this has been analyzed in journalistic spheres on and off the Internet, what hasn't been discussed is that a main objective of this particular Zionist false flag was to drive a wedge between Pakistan and Iran and foment suspicion between them. Thankfully, despite the terrible, terrible loss of Iranian life, the brotherly Islamic Republics have gotten over this and are moving forward with an important gas project in the face of immense pressure from the Zionist-occupied United States government.

William Safire, infamous neconservative ideologue and member of the Jewish-Zionist brain trust that planned the annihilation of Iraq, was well-known for supporting the formation of a Kurdish state because it would weaken the Zionist entity's three greatest enemies, Iraq, Iran and Syria. This exact line of Zionist thinking is present in the minds of Safire's constituents at the powerful neoconservative think tank, the Brookings Institution, which produced a policy paper entitled, "Which Path To Persia?" that lays out numerous Israelcentric strategies for America and "Israel" to destroy Iran, including the launch of an insurgency in Balochistan with Pakistan being transformed into a total slave-state to support the terrorism against Iran by proxy. "Which Path To Persia?" was authored by notorious 'Iranophobe' and Jewish supremacist Kenneth M. Pollack, world-renowned lobbyist for the usurping Zionist regime Martin Indyk, Jewish-Zionist CIA stalwart and developer of Obama's Af-Pak strategy Bruce Reidel, Jewish supremacist defender of the criminal "Israeli" practice of "targeted killings" Daniel L. Byman and neoconservative shabbos goyimSuzanne Maloney and Michael E. O'Hanlon.

"Israeli" foreign policy advisor Oded Yinon's "A Strategy For Israel In The 1980s", the monumental 1982 paper that documented in detail how the Jewish occupation regime plans to balkanize its enemies through infiltration, subversion and terrorism, rendering them non-existent as robust nation-states and turning them into satellites of Zionist hegemony, was instrumental in the Jewish decimation of Iraq, and it is now being implemented in numerous nations across the continent of Africa. The Yinon plan is also very much underway in Pakistan, with the ultimate territorial goal of the Zionist entity being to break off Balochistan from the Islamic Republic, as well as a larger chunk of land in the western part of the country to make way for a gutted Afghanistan called "Pashtunistan," thus crippling Pakistan in a devastating way, essentially reincarnating and modernizing the "Agartala Conspiracy" of 1971.

The convergence of Zionist-Hindutvadi objectives in Balochistan is by no means a coincidence. It is in fact exceedingly deliberate. A sister of the Dragon Policy, the Yinon-styled infiltration of Balochistan is a plot more than 10 years old known as "Operation Blue Tulsi." The resurrection of the BLA and its offshoots, the training camps set up for them, the bases set up by Mossad, RAW and the occasional cluster of CIA agents to run the terror squads, the BLA designs to overthrow the Islamabad government in 2004, the setup of a Baloch regime office in occupied al-Quds to better coordinate Mossad-RAW-CIA subversion schemes, the assassinations of top military and intelligence officials followed by their replacement with infiltrators from the Zionist-Hindutvadi nexus, and all other facsimile events are part and parcel of Operation Blue Tulsi, which is being executed simultaneously alongside Cold Start and Operation Kurukshetra.

On a side note, it is interesting to record that the "Pakistani" troops exported to Bahrain to be used in the crackdown on a revolutionary uprising in that country at the behest of its ruling regime, the criminal and barbaric Al-Khalifa monarchy, which has bragged about its ties to Mossad and has told all of its top officials not to refer to "Israel" as the "enemy" or the "Zionist entity," are not from the nationalistic and anti-Zionist military-intelligence apparatus of Pakistan, but from Balochistan. Not only is Mossad-RAW-backed terrorism from Balochistan wreaking havoc on the lives of innocent Pakistanis and Kashmiris, but Bahrainis too.

With revelations having emerged from Jewish former US Ambassador to India John Gunther Dean that Mossad and RAW orchestrated the "accident" that murdered General Muhammad Zia-ul-Haq in 1988, it seems that there isn't a single segment of the tragic aspects of Pakistan's tumultuous history that Zionism and Hindutva didn't have their dirty hands in. Moreover, the Zionist-controlled United States has started bombing the "child of Jinnah, child of 'Allama Iqbal" once again with drones as sadistic revenge for cutting off NATO supplies to occupied Afghanistan ; the latest attack has left four little Pakistani girls slaughtered. And with the egregious and monstrous human rights abuses in occupied Kashmir continuing en masse, as most recently evidenced by the Indian occupation army being implicated in punishing Kashmiris through slave labor, and the more than 1,332 Kashmiris locked up in Hindutvadi dungeons through the draconian PSA over the last three years, the aforesaid plans for another intifada in the Vale couldn't have come at a better time. The filthy and pernicious plots of the Zionist-Hindutvadi alliance against the peoples of Pakistan and occupied Kashmir must be confronted and stopped once and for all.

To reiterate, it indeed is not an exaggeration to suggest that Pakistan, occupied Kashmir, and, due to it being included in the enemy plots against these nation-states, the Iranian Islamic Revolution, stand on the brink of oblivion. The usurping Jewish entity and the Hindutvadi New Delhi regime are doped up on geopolitical and religious fundamentalism that is driving each and every decision that they make in pursuit of two separate empires: Greater "Israel", the Zionist confiscation and domination of all Arab lands between the Nile and the Euphrates, and Hindutva, the Hindutvadi confiscation and domination of every piece of territory within the Southeast Asia Subcontinent. Whoever stands in their way faces the unfortunately but nevertheless extremely real prospect of a ruthless militaristic backlash at the hands of these two supremacist allies. It is not only because of the "Israeli"-Indian drive for hegemony that Pakistan, Kashmir and Iran are being targeted, it is because of their defiance of their enemies' thieving thirst for Muslim land and treasure that they are in the reticle; they refuse to be subjugated, they refuse to bow before anyone except their Creator.

Pakistan, Kashmir and Iran, along with the Resistance parties in Syria, Lebanon, occupied Palestine, Bahrain, Iraq, Somalia, Belarus, Yemen, Bolivia, Venezuela and anywhere else in the world where there is struggle in the shadows against this mad and bloodthirsty nexus of supremacism and thuggery are humanity's last hope for a better world. A just world. A peaceful world. A world of tolerance and understanding. A world of brotherhood and sisterhood, free from internationalist Jewish usury and the ever-looming threat of Zionist-Hindutvadi expansionism. Such a world is possible and the aforementioned "Peoples of Resistance" show us that it is so day in and day out. May their miraculous defiance persist in inspiring us all and spread to every corner of the globe until Zionism, Hindutva and all of the marionettes attached to their machinations fade away into the most damned pages of history. If it does, Azadi for everyone is just around the corner.

THROUGH A PAKISTANI LENS

NEARLY four decades after the secession of East Pakistan – now Bangladesh – most Pakistanis continue to see Bangladesh and its experiments with democracy through a three-faceted prism. The initial feeling of acrimony over the dismemberment of their nation, or the hope that one day Pakistan and Bangladesh will either re-unite or form a confederation, are today merely part of a faded discourse that no longer enjoysCURRENCY in public or official sentiments. The psyche of both the Pakistani people and the state over the wound caused by the secession of Bangladesh seems to have healed itself far quicker than has ostensibly happened in India over the separation of Pakistan where, apparently, a perspective clouded by resentment still resonates, even if it's not dominant.

In fact, our own obsession with India and an India-centric approach remains the predominant lens through which we look at Bangladesh. One major approach is based on our understanding of Bangladesh's relations with India. Clearly the fact that India is the largest exporter to Bangladesh creates substantial unease in Pakistan. Many Pakistanis continue to characterize political parties in Bangladesh as being either 'soft' or 'tough' on India. The overwhelming victory of the Awami League and its allies, who won an impressive 263 seats in the 300 member Parliament in December 2008, continues to be a cause of anxiety in Pakistan given the pro-India image of the Awami League. The Awami League trounced its rival Bangladesh Nationalist Party (BNP)-Jamaat-e-Islami coalition, which managed only 32 seats.

Tensions between Bangladesh and India appear to please many in Pakistan and closer relations between the two countries are viewed with suspicion. Irrational or jingoistic but such is the sad reality which both media and intellectuals reflect in their occasional public, but generally private, discussions. Against this backdrop, the Awami League victory has not been perceived as a source of great satisfaction in Pakistan. The sentiment among certain segments of the Bangladeshi political elite that the Bangladesh 'establishment' sided with the Awami League to present a 'liberal face' of Bangladesh to the world is echoed in Pakistan's discussion rooms. Unfortunately, after a rather promising start, the current state of relations between Pakistan and India have once again become strained following the terrorist attack in Mumbai. One hopes that as relations between the two countries normalize, this habit of viewing Bangladesh's politics through an Indian angle will subside.

A second prism through which we in Pakistan view Bangladesh revolves around the role of its military in politics. Expectedly, given our own troubled history with military intervention in politics, the role of the Bangladesh military in the country's politics generates substantial interest in Pakistan.

Quite unlike the decade of the 1990s when, despite weaknesses in the political process, it appeared that Pakistan was gradually moving towards a democratic consolidation, the period between 1975 and 1991 in Bangladesh was dominated by its army, a development very similar to what has usually happened in Pakistan.

Yet, instead of welcoming the change, the democratic transfer of power to a civilian government in Bangladesh in 1991 created some consternation in Pakistan. The dominant view, shared with leading Bangladeshi scholars, was that since the Bangladesh military was happily engaged in peacekeeping missions overseas, it would not risk the ire of the international community by intervening in the internal politics of the country.

This understanding, however, had to be revised as the powerful Bangladesh military intervened in politics once again in 2007. Fortunately, however, it stopped short of turning the intervention into a full-scale coup and instead facilitated the appointment of a dozen businessmen, technocrats and former diplomats to manage the administration.

A key reason why the Bangladesh Army preferred to stay behind the scenes was that it did not want to lose out on the opportunity to participate in UN peacekeeping missions, as it accounted for a considerable sum ofMONEY coming into Bangladesh, directly benefiting army personnel. Many in Pakistan wonder whether such a reward system could similarly work in their own case.

Yet while many in Pakistan give credit to the inventive model of the military coup in Bangladesh, they are less sanguine about its applicability in their own country, especially when the 18 month itch with a democratically elected government has already kicked in and stories of alleged corruption and incapacity for good governance are rather widespread.

Nevertheless, having tried the Ayub Khan, Zia-ul-Haq and Musharraf models – all hands-on, very intrusive and boasting of 'unity of command' – the 'behind-the-scene', 'reform-oriented' role of the Bangladesh military in 2007-08, backed by the silent or not-so-silent cheers of international diplomats, looks just too tempting to the democracy-fickle millions in Pakistan. Bangladesh's impressive electoral reforms, that we discuss ahead, are believed to have been achieved under the military's watchful eye, an accomplishment that eight years of the Musharraf regime could not boast of in Pakistan.

Bangladesh witnessed a peaceful transition of power from a military-backed caretaker government to a democratically elected government in 2008. The successful handling of the ninth parliamentary election in December 2008 by the military, and the wide scale of political and electoral reforms preceding the election, has been viewed with admiration in Pakistan. When it comes to electoral reforms, many feel that Pakistan's political leadership has a few lessons to learn from the country's former eastern wing that chose to go its separate way and became the People's Republic of Bangladesh in 1971.

Although Bangladesh was a late starter on electoral reforms, it has made quick and impressive strides contributing to a smooth election and subsequent quick transfer of power. The Bangladesh Election Commission (BEC) became a truly independent body only during the past couple of years as a result of the intensive electoral reforms undertaken by the caretaker government. These reforms became possible only after the government appointed a Chief Election Commissioner (CEC) who was not only considered reform-minded and dynamic, but had impeccable administrative experience.

The extensive electoral reforms were the result of a comprehensive dialogue between the BEC and the political parties. As many as sixteen political parties were involved in three rounds of dialogue. In addition, media and civil society were also consulted by the BEC on the reforms before they were introduced in the form of various laws. In contrast, dialogue between the Election Commission and political parties is rare in Pakistan. The last time the Elections Commission of Pakistan consulted political parties ahead of the general elections was in 2008 for merely a couple of hours and that too at the direction of the Supreme Court.

Pakistani observers viewed the preparation of an accurate voters list in Bangladesh with considerable envy. The Bangladesh Election Commission (BEC), with the active participation of the army, was able to prepare a voter list of about 81 million people which is digitised, carries each voter's photograph, includes close to 95% of all eligible voters and, above all, has won the confidence of all political parties, citizens' organizations and the international community. The list took just eleven months to prepare.

A significant bye-product of the voter list is a National Identification Card for the adult (18 years and above) population of Bangladesh. Photographs of all adults, including women, were mandatory, for both the voter list and identification card. By integrating each voter's (including those of *purdah*-observing women) picture into the voters list, the commission has satisfactorily solved the problem of voter identification. Voter identification remains a major challenge, creating perpetual controversy as a potential source of bogus vote casting, especially at women's polling stations in Pakistan, Bangladesh and, to some extent, India. Moreover, incorporating the voters' picture in the electoral rolls has precluded the need of carrying any additional identification document by voters.

The Election Commission of Pakistan too had commissioned similar computerised electoral rolls way back in April 2006, and completed the exercise in about 20 months, just a few weeks before the 2008 general election. Unfortunately, the so-called computerised electoral rolls of Pakistan neither contain the individual's picture, nor are they complete or error free. Despite a door-to-door survey and an expense of over a billion rupees, the resulting voter list does not generate requisite confidence in either the political parties or citizens' groups. Our list is replete with errors, viz. multiple entries of voters, and thus a major source of dissatisfaction for various candidates in the 2008 election.

Further, despite the fact that Pakistan instituted a system of National Identification Cards as far back as 1974, and adopted computerised identification cards through the National Database and Registration Authority (NADRA) in 1998, even today nearly a quarter of the adult population (around 25%) remains without identification cards. The Bangladesh government, by contrast has performed the feat of providing identification cards to over 95% of its adult population in less than a year. The process of voter registration and preparation of a National Identification Card is an integrated process in Bangladesh, eliminating time-consuming duplication of effort by citizens and state institutions.

In addition to the success story of the compilation of electoral rolls, the recent elections in Bangladesh set an admirable example of fair and orderly election. As any politician with experience of electoral politics will testify, a major expense of candidates is incurred on mobilising transport for voters on polling day. In order to minimize the influence ofMONEY in elections, the Bangladesh Election Commission prohibited the use of all motorised vehicles on election day, except for inter-city traffic and emergency use. No banners or hoardings were allowed by the BEC either. Candidates were allowed to print only black and white posters of a certain size, and attach them to strings that were hung across the streets. Posters could not be pasted on any wall or structure. Painting on walls was not permitted. All these instructions were strictly followed and none of the international observers found any violations. The result was clean walls, resulting in huge savings on cleaning and re-painting. Another interesting electoral reform was the 30-page exhaustive disclosure and declaration form that each candidate was required to file along with nomination papers. The disclosures and declarations were made public by the Bangladesh Election Commission and placed on its website for public scrutiny. Each candidate was required to state his/her educational qualification; details of any pending criminal cases; any outstanding amount payable to any state institution; details of any outstanding or written-off loans payable to any bank orFINANCIAL institution, and so on. Whether or not such public disclosures can be made mandatory in Pakistan, it is nevertheless clear that such disclosures help voters decide about the 'suitability' of candidates.

Another innovative feature of the 2008 parliamentary election in Bangladesh was the inclusion of a 'no vote' provision for voters unhappy with all candidates. In the 2008 election, some 380,000 voters cast a 'no vote', accounting for about 0.5% of the total votes cast. The election rules provide that if over 50% of the voters chose to cast a no vote in a particular constituency, it would automatically lead to re-election. Fortunately, no constituency faced this scenario in the 2008 election.

News about the living conditions of stranded 'Bihari Pakistanis' in Bangladesh is prominently carried in Pakistan. This time, however, a sizeable number of the Urdu speaking population, who had been living in camps for the last 38 years and consistently refusing to accept Bangladeshi citizenship by claiming to be Pakistanis, opted to register as voters in the 2008 election after becoming Bangladeshi nationals. Although their exact number is not available, it is widely accepted that most of the Urdu-speaking voters did vote in this election. Candidates in constituencies with a sizeable number of Urdu-speaking voters even published election posters in Urdu, a fact highlighted by Bangladeshi media. Interestingly, as many as 87% of registered Bangladeshi voters turned out to vote in the 2008 election. Pakistan, by contrast, suffers from a chronic and embarrassingly low voter turnout, the lowest average turnout among countries of South Asia, and one of the lowest in the world. (Pakistan ranks 164 among 169 countries studied by International Institute for Democracy and Electoral Assistance – IIDEA – since 1945.)

After the seemingly successful intrusion of the military into politics and delivering a rather admirable election to the people of Bangladesh in December 2008, many in Pakistan forecast that the army chief, General Moeen U. Ahmed would assume the presidency of the country. They were proved wrong. The General retired and a political figure was elected as the new President of Bangladesh.

Many Pakistanis also expected that following a smooth and orderly election, politics in Bangladesh would become more stable and less fractious. However, this assumption appears to be somewhat premature. Bangladesh seems to be following in the footsteps of Pakistan's fast-fading environment of 'reconciliation' between Nawaz Sharif's PML-N and the Zardari led PPP. Even though Sheikh Hasina has promised to set a new example in politics in line with her promised 'charter of change' by seeking opposition support in decision-making and even giving Khaleda Zia, the former prime minister, a position in the government, reconciliation and cooperation between the two parties remains somewhat distant.

Nawaz Sharif's party and the ruling PPP had locked horns throughout the decade of the 1990s, much like the acrimonious relationship that exists between the two leading political parties in Bangladesh. The signing of a Charter of Democracy in May 2006 between the late Benazir Bhutto and Nawaz Sharif did raise hopes that the two parties were willing to learn from the past. Yet, while the charter of democracy does seem to have guided the PPP and PML-N in forming a coalition government after the February 2008 elections in Pakistan, the 'conciliatory' atmosphere did not last long. Fortunately though, the PML-N, which occupies the opposition benches in the National Assembly today, has not let its opposition to the government reach acrimonious levels. The government-opposition relationship in Bangladesh, in contrast, seems to be much more tense.

Although the Sheikh Hasina government has yet to make major headway in rooting out corruption or containing militancy, she remains remarkably popular in the country with 78% people very satisfied or satisfied with her job performance. In contrast, the popularity rating of President Zardari of Pakistan stands at a low 32%, down from 64% a year ago. Another poll released by Gallup International in July 2009 records public confidence in the Bangladesh government at 87% while 88% approve the job performance of Prime Minister Sheikh Hasina. The performance rating of the Bangladesh leader of the opposition, Khaleda Zia, at 41% is quite a contrast from 79% popularity ratings of Nawaz Sharif in Pakistan, head of the leading opposition party, Pakistan Muslim League-Nawaz (PML-N). Pakistan's Prime Minister Gilani, a stalwart of the PPP, hand-picked by President Zardari, incidentally enjoys 67% approval ratings.

Economic progress is another point of reference for viewing Bangladesh from Pakistan. Prior to East Pakistan seceding and becoming Bangladesh, a small but influential section of the Pakistani establishment unjustifiably believed that East Pakistan was a drag on the national economy and that Pakistan would develop much faster after the break-up. On the other hand, many people in the then eastern wing of the country believed that the western wing was exploiting their resources and that they had been deprived of their rightful share in national development allocations.

Nearly four decades later, the people of Pakistan are naturally curious about how the estranged former eastern wing has fared on the economic front. Development professionals and economists in Pakistan are envious of Bangladesh's success at controlling its high rate of population growth. At the time of independence, Bangladesh was more populous than its erstwhile western wing. This 'superiority' in numbers was not recognised by the western wing for most part of their journey together from 1947 to 1971, as the first Constitution of the country developed the principle of 'parity' between the two wings by allocating an equal number of parliamentary seats to both despite the numerical superiority of the eastern wing. The Bangladesh population, at around 140 million in 2006, is growing at an annual rate of about 1.4% compared to around 3% in the period up to 1990. In contrast, the Pakistan population grew at around 2.6% per annum, a figure that has only now come down to 1.8% per annum. Its population, as of 2006, now stands at 157 million.

Despite a higher population, Pakistan exports fewer workers to international MARKETS as compared to Bangladesh. As of 2007, Bangladeshi workers around the world numbered 981,000 compared to around 300,000 Pakistani overseas workers. As a result Bangladesh was able to earn close to US $8,000 million in remittances through its overseas workers in 2007-08 compared to about US $6,451 million by Pakistani migrant workers. Pakistan is envious of the growing competitiveness of Bangladeshi workers in the international MARKET.

Pakistanis also note that Bangladesh's GDP growth rate during the past decade or so has been impressive, at times higher than that of Pakistan. For example, the GDP growth rate in Bangladesh was 6.21% in 2007-08 compared to Pakistan's 4.1% in the same period. Although total Pakistani exports are still higher, Bangladeshi exports of cotton garments and finished cotton goods are a source of envy for Pakistan, especially as it is one of the largest growers of cotton whereas Bangladesh virtually has to import all its raw cotton for its finished cotton goods. Bangladeshi exports of finished cotton goods worth over US $10,000 million compared to Pakistan's US $7,500 million during the year 2007-08, illustrates the point.

Overall then, close to four decades since the coming into being of Bangladesh, Pakistanis continue to view the country's political and economic developments through the prism of their own mixed track record.

9

R&AW and Baluchistan

As the purported assassination of Osama bin Laden has placed the focus on Pakistan, it is vital to assess the changing role of Pakistan in broad geostrategic terms, and in particular, of the changing American strategy toward Pakistan. The recently reported assassination was a propaganda ploy aimed at targeting Pakistan. To understand this, it is necessary to examine how America has, in recent years, altered its strategy in Pakistan in the direction of destabilization. In short, Pakistan is an American target. The reason: Pakistan's growing military and strategic ties to China, America's primary global strategic rival. In the 'Great Game' for global hegemony, any country that impedes America's world primacy – even one as historically significant to America as Pakistan – may be sacrificed upon the altar of war.

'Pakistan in Pieces' examines the changing views of the American strategic community – particularly the military and intelligence circles – towards Pakistan. In particular, there is a general acknowledgement that Pakistan will very likely continue to be destabilized and ultimately collapse. What is not mentioned in these assessments, however, is the role of the military and intelligence communities in making this a reality; a veritable self-fulfilling prophecy. This part also examines the active on the ground changes in American strategy in Pakistan, with increasing military incursions into the country.

IMPERIAL EYE ON PAKISTAN

In December of 2000, the CIA released a report of global

trends to the year 2015, which stated that by 2015, "Pakistan will be more fractious, isolated, and dependent on international financial assistance." Further, it was predicted, Pakistan:

Will not recover easily from decades of political and economic mismanagement, divisive politics, lawlessness, corruption and ethnic friction. Nascent democratic reforms will produce little change in the face of opposition from an entrenched political elite and radical Islamic parties. Further domestic decline would benefit Islamic political activists, who may significantly increase their role in national politics and alter the makeup and cohesion of the military – once Pakistan's most capable institution. In a climate of continuing domestic turmoil, the central government's control probably will be reduced to the Punjabi heartland and the economic hub of Karachi.

The report further analyzed the trends developing in relation to the Pakistan-India standoff in the region:

The threat of major conflict between India and Pakistan will overshadow all other regional issues during the next 15 years. Continued turmoil in Afghanistan and Pakistan will spill over into Kashmir and other areas of the subcontinent, prompting Indian leaders to take more aggressive preemptive and retaliatory actions. India's conventional military advantage over Pakistan will widen as a result of New Delhi's superior economic position.

In 2005, the Times of India reported on a US National Intelligence Council report, written in conjunction with the CIA, which predicted a "Yugoslavia-like fate" for Pakistan, saying that, "by year 2015 Pakistan would be a failed state, ripe with civil war, bloodshed, inter-provincial rivalries and a struggle for control of its nuclear weapons and complete Talibanisation."

In November of 2008, the US National Intelligence Council released a report, "Global Trends 2025," in which they outlined major trends in the world by the year 2025. When it came to Pakistan, the report stated that, "Ongoing low-intensity clashes between India and Pakistan continue to raise the specter that such events could escalate to a broader conflict between those nuclear powers." It stated that Pakistan "will be at risk of state failure." In examining potential failed states, the report stated that:

Youth bulges, deeply rooted conflicts, and limited economic prospects are likely to keep Palestine, Yemen, Afghanistan, Pakistan, and others in the high-risk category. Spillover from turmoil in these states and potentially others increases the chance that moves elsewhere in the region toward greater prosperity and political stability will be rocky.

The report referred to Pakistan as a "wildcard" and stated that if it is "unable to hold together until 2025, a broader coalescence of Pashtun tribes is likely to emerge and act together to erase the Durand Line [separating Pakistan from Afghanistan], maximizing Pashtun space at the expense of Punjabis in Pakistan and Tajiks and others in Afghanistan."

In January of 2009, a Pentagon report analyzing geopolitical trends of significance to the US military over the next 25 years, reported that Pakistan could face a "rapid and sudden" collapse. It stated that, "Some forms of collapse in Pakistan would carry with it the likelihood of a sustained violent and bloody civil and sectarian war, an even bigger haven for violent extremists, and the question of what would happen to its nuclear weapons," and as such, "that 'perfect storm' of uncertainty alone might require the engagement of U.S. and coalition forces into a situation of immense complexity and danger."

A top adviser to former President George Bush and current President Obama warned in April of 2009, that Pakistan could collapse within months, and that, "We have to face the fact that if Pakistan collapses it will dwarf anything we have seen so far in whatever we're calling the war on terror now." The adviser and consultant, David Kilcullen, explained that this would be unlike the conflicts in Afghanistan and Iraq, which each had a population of over 30 million, whereas "Pakistan has [187] million people and 100 nuclear weapons, an army which is bigger than the American army, and the headquarters of al-Qaeda sitting in two-thirds of the country which the Government does not control."

TARGET: PAKISTAN

Going back to the later years of the Bush administration, it is apparent that the US strategy in Pakistan was already changing in seeing it increasingly as a target for military operations as opposed to simply a conduit. In August of 2007, newly uncovered documents revealed that the US military "gave elite units broad authority" in 2004, "to pursue suspected terrorists into Pakistan, with no mention of telling the Pakistanis in advance."

In November of 2007, an op-ed in the New York Times stated categorically that, "the United States simply could not stand by as a nuclear-armed Pakistan descended into the abyss," and that, "we need to think — now — about our feasible military options in Pakistan, should it really come to that." The authors, Frederick Kagan and Michael O'Hanlon are both well-known strategists and scholars at the American Enterprise Institute and Brookings Institution, two of the most prominent and influential think tanks in the United States. While stating that Pakistan's leaders are still primarily moderate and friendly to the US, "Americans felt similarly about the shah's regime in Iran until it was too late," referring to the outbreak of the Iranian Revolution in 1979. They warn:

The most likely possible dangers are these: a complete collapse of Pakistani government rule that allows an extreme Islamist movement to fill the vacuum; a total loss of federal control over outlying provinces, which splinter along ethnic and tribal lines; or a struggle within the Pakistani military in which the minority sympathetic to the Taliban and Al Qaeda try to establish Pakistan as a state sponsor of terrorism.

They state that the military solutions are "daunting" as Pakistan is a nation of 187 million people, roughly five times the size of Iraq. They wrote that, "estimates suggest that a force of more than a million troops would be required for a country of this size," which led them to conclude, "Thus, if we have any hope of success, we would have to act before a complete government collapse, and we would need the cooperation of moderate Pakistani forces." They suggested one plan would be to deploy Special Forces "with the limited goal of preventing Pakistan's nuclear materials and warheads from getting into the wrong hand." However, they admit that, "even pro-American Pakistanis would be unlikely to cooperate." Another option, they contend: would involve supporting the core of the Pakistani armed forces as they sought to hold the country together in the face of an ineffective government, seceding border regions and Al Qaeda and Taliban assassination attempts against the leadership. This would require a sizable combat force – not only from the United States, but ideally also other Western powers and moderate Muslim nations.

The authors concluded, saying that any state decline in Pakistan would likely be gradual, therefore allowing the US to have time to respond, and placed an emphasis on securing Pakistan's nuclear arsenal and combating militants. They finished the article with the warning: "Pakistan may be the next big test."

In December of 2007, the Asia Times Online ran a story about the US plan to rid Pakistan of President Musharraf, and that the US and the West, more broadly, had begun a strategy aimed at toppling Pakistan's military. As part of this, the US launched a media campaign aimed at demonizing Pakistan's military establishment. At this time, Benazir Bhutto was criticizing the ISI, suggesting they needed a dramatic restructuring, and at the same time, reports were appearing in the US media blaming the ISI for funding and providing assistance to Al-Qaeda and the Taliban. While much of this is documented, the fact that it suddenly emerged as talking points with several western officials and in the media does suggest a turn-around against a long-time ally.

Both Democratic and Republican politicians were making statements that Pakistan represented a greater threat than Iran, and then-Senator (now Vice President) Joseph Biden suggested that the United States needed to put soldiers on the ground in Pakistan in cooperation with the "international community." Biden said that, "We should be in there," and "we should be supplying tens of millions of dollars to build new schools to compete with the madrassas. We should be in there building democratic institutions. We should be in there, and get the rest of the world in there, giving some structure to the emergence of, hopefully, the reemergence of a democratic process."

In American policy-strategy circles, officials openly began discussing the possibility of Pakistan breaking up into smaller states, and increasing discussion that Musharraf was going to be "removed," which obviously happened. As the Asia Times stated:

Another worrying thing is how US officials are publicly signaling to the Pakistanis that Bhutto has their backing as the next leader of the country. Such signals from Washington are not only a kiss of death for any public leader in Pakistan, but the Americans also know that their actions are inviting potential assassins to target Bhutto.

If she is killed in this way, there won't be enough time to find the real culprit, but what's certain is that unprecedented international pressure will be placed on Islamabad while everyone will use their local assets to create maximum internal chaos in the country.

Of course, this subsequently happened in Pakistan. As the author of the article pointed out with startlingly accurate foresight, "Getting Bhutto killed can generate the kind of pressure that could result in permanently putting the Pakistani military on a back foot, giving Washington enough room to push for installing a new pliant leadership in Islamabad." He observed that, "the US is very serious this time. They cannot let Pakistan get out of their hands."

Thus, it would appear that the new US strategic aim in Pakistan was focused on removing the Pakistani military from power, implying the need to replace Musharraf, and replace him with a new, compliant civilian leadership. This would have the effect of fracturing the Pakistani elite, threatening the Army's influence within Pakistani politics, and undertaking more direct control of Pakistan's government.

As if on cue, in late December it was reported that, "US special forces snatch squads are on standby to seize or disable Pakistan's nuclear arsenal in the event of a collapse of government authority or the outbreak of civil war following the assassination of Benazir Bhutto."

The New York Times ran an article in early January 2008, which reported that, "President Bush's senior national security advisers are debating whether to expand the authority of the Central Intelligence Agency and the military to conduct far more aggressive covert operations in the tribal areas of Pakistan." The article stated that the new strategy was purportedly in response to increased reports of Al-Qaeda and Taliban activity within

Pakistan, which "are intensifying efforts there to destabilize the Pakistani government." Bush's National Security team supposedly organized this effort in response to Bhutto's assassination 10 days previously.

Officials involved in the strategy discussions said that some "options would probably involve the C.I.A. working with the military's Special Operations forces," and one official said, "After years of focusing on Afghanistan, we think the extremists now see a chance for the big prize — creating chaos in Pakistan itself." Of pivotal importance to the strategy, as the Times reported: "Critics said more direct American military action would be ineffective, anger the Pakistani Army and increase support for the militants." Perhaps this is not simply a "side-effect" of the proposed strategy, but in fact, part of the strategy.

As one prominent Pakistani political and military analyst pointed out, raids into Pakistan would expand anger and "prompt a powerful popular backlash" against the Pakistani government, losing popular support. However, as I previously stated, this might be the intention, as this would ultimately make the government more dependent upon the United States, and thus, more subservient. On September 3, 2008, it was reported that a commando raid by US Special Forces was launched in Pakistan, which killed between 15 and 20 people, including women and children. The Special Forces were accompanied by five U.S. helicopters for the duration of the operation.

In February of 2009, it was reported that, "More than 70 United States military advisers and technical specialists are secretly working in Pakistan to help its armed forces battle Al Qaeda and the Taliban in the country's lawless tribal areas." So not only are U.S. Special Forces invading Pakistani territory; but now US military advisers are secretly advising the Pakistani Army on its own operations, and the advisers are themselves primary made up of Special Forces soldiers. They provide the Pakistani Army "with intelligence and advising on combat tactics," and make up a secret command run by US Central Command and Special Operations Command (presumably JSOC – Joint Special Operations Command).

In May of 2009, it was reported that, "the U.S. is sending Special Forces teams into one of Pakistan's most violent regions as part of a push to accelerate the training of the Pakistani military and make it a more effective ally in the fight against insurgents there." The Special Forces were deploying to two training camps in the province of Baluchistan, and "will focus on training Pakistan's Frontier Corps, a paramilitary force responsible for battling the Taliban and al Qaeda fighters." Further, the project "is a joint effort with the U.K.," which helps "fund the training, although it is unclear if British military personnel would take part in the initiative. British officials have been pushing for such an effort for several years."

In December of 2009 it was revealed that, "American special forces have conducted multiple clandestine raids into Pakistan's tribal areas as part of a secret war in the border region where Washington is pressing to expand its drone assassination programme," which was revealed by a former NATO officer. He said these incursions had occurred between 2003 and 2008, indicating they go even further back than US military documents stipulate. The source further revealed that, "the Pakistanis were kept entirely in the dark about it. It was one of those things we wouldn't confirm officially with them." Further, as the source noted, British "SAS soldiers have been active in the province" of Bolochistan in 2002 and 2003 and "possibly beyond."

The "Balkanization" of Pakistan: Blaming the Pakistanis

Selig S. Harrison is a director of the Asia Program at the Center for International Policy, senior scholar of the Woodrow Wilson International Center for Scholars, former senior associate of the Carnegie Endowment for International Peace, and former journalist and correspondent. "His reputation for giving 'early warning' of foreign policy crises was well established during his career as a foreign correspondent. In his study of foreign reporting, Between Two Worlds, John Hohenberg, former secretary of the Pulitzer Prize Board, cited Harrison's prediction of the 1965 Indo-Pakistan war eighteen months before it happened." Further, "More than a year before the Russians invaded Afghanistan, Harrison warned of this possibility in one of his frequent contributions to the influential journal Foreign Policy." On February 1, 2008, Selig Harrison threw his renowned "predictive" abilities on Pakistan in an op-ed for the New York Times in the run-up to the Pakistani elections. He started by stating that, "Whatever the outcome of the Pakistani elections, now scheduled for Feb. 18, the existing multiethnic Pakistani state is not likely to survive for long unless it is radically restructured." Harrison then went on to explain that Pakistan would likely break up along ethnic lines; with the Pashtuns, concentrated in the northwestern tribal areas, the Sindhis in the southeast uniting with the Baluch tribesmen in the southwest, with the Punjab "rump state" of Pakistan.

The Pashtuns in the north, "would join with their ethnic brethren across the Afghan border (some 40 million of them combined) to form an independent 'Pashtunistan'," and the Sindhis "numbering 23 million, would unite with the six million Baluch tribesmen in the southwest to establish a federation along the Arabian Sea from India to Iran," presumably named Baluchistan; while the rump state of Pakistan would remain Punjabi dominated and in control of the nuclear weapons. Selig Harrison explained that prior to partition from India, which led to the creation of the Pakistani state in 1947, Pashtun, Sindhi and Baluch ethnicities had "resist Punjabi domination for centuries," and suddenly:

> they found themselves subjected to Punjabi-dominated military regimes that have appropriated many of the natural resources in the minority provinces — particularly the natural gas deposits in the Baluch areas — and siphoned off much of the Indus River's waters as they flow through the Punjab.
>
> The resulting Punjabi-Pashtun animosity helps explain why the United States is failing to get effective Pakistani cooperation in fighting terrorists. The Pashtuns living along the Afghan border are happy to give sanctuary from Punjabi forces to the Taliban, which is composed primarily of fellow Pashtuns, and to its Qaeda friends.
>
> Pashtun civilian casualties resulting from Pakistani and American air strikes on both sides of the border are breeding a potent underground Pashtun nationalist movement. Its initial objective is to unite all Pashtuns in Pakistan, now divided among

political jurisdictions, into a unified province. In time, however, its leaders envisage full nationhood.

... The Baluch people, for their part, have been waging intermittent insurgencies since their forced incorporation into Pakistan in 1947. In the current warfare Pakistani forces are widely reported to be deploying American-supplied aircraft and intelligence equipment that was intended for use in Afghan border areas. Their victims are forging military links with Sindhi nationalist groups that have been galvanized into action by the death of Benazir Bhutto, a Sindhi hero as was her father, Zulfikar Ali Bhutto.

This passage is very revealing of the processes and perceptions surrounding "Balkanization" and "destabilization." What I mean by this, is that historically and presently, imperial powers would often use ethnic groups against each other in a strategy of divide and conquer, in order "to keep the barbarians from coming together" and dominate the region.

Zbigniew Brzezinski wrote in his 1997 book, "The Grand Chessboard," that, "Geopolitics has moved from the regional to the global dimension, with preponderance over the entire Eurasian continent serving as the central basis for global primacy." Brzezinski then gave a masterful explanation of the American global strategy, which placed it into a firm imperialistic context:

To put it in a terminology that hearkens back to the more brutal age of ancient empires, the three grand imperatives of imperial geostrategy are to prevent collusion and maintain security dependence among the vassals, to keep tributaries pliant and protected, and to keep the barbarians from coming together.

While imperial powers manipulate, and historically, even create the ethnic groups within regions and nations, the West portrays conflict in such regions as being the product of these "ethnic" or "tribal" rivalries. This perception of the East (Asia and the Middle East) as well as Africa is referred to as Orientalism or Eurocentrism: meaning it generally portrays the East (and/or Africa) as "the Other": inherently different and often barbaric. This prejudiced perspective is prevalent in Western academic, media, and policy circles. This perspective serves a major purpose: dehumanizing a people in a region that an imperial power seeks to dominate, which allows the hegemon to manipulate the people and divide them against each other, while framing them as "backwards" and "barbaric," which in turn, justifies the Western imperial power exerting hegemony and control over the region; to "protect" the people from themselves.

Historically and presently, Western empires have divided people against each other, blamed the resulting conflict on the people themselves, and thus justified their control over both the people, and the region they occupy. This was the strategy employed in major recent geopolitical conflicts such as the breakup of Yugoslavia and the Rwandan genocide. In both cases, Western imperial ambitions were met through exacerbating ethnic rivalries, providing financial, technical, and military aid and training to various factions; thus, spreading violent conflict, war, and genocide. In both cases, Western, and primarily American strategic interests were met through an increased presence militarily, pushing out other major imperial and powerful rivals, as well as increasing Western access to key economics resources.

This is the lens through which we must view the unfolding situation in Pakistan. However, the situation in Pakistan presents a far greater potential for conflict and devastation than either Yugoslavia or Rwanda. In short, the potential strategy of "Balkanization" and destabilization of Pakistan could dwarf any major global conflict in the past few decades. It's sheer population of 187 million people, proximity to two major regional wars in Iraq and Afghanistan, its strategic location as neighbor to India, China, and Iran with access to the Indian Ocean, and its nuclear arsenal, combine to make Pakistan the potential trigger for a much wider regional and possibly global war. The destabilization of Pakistan has the potential to be the greatest geopolitical catastrophe since World War II.

Thus, Selig Harrison's op-ed in the New York Times in which he describes the "likely" breakup of Pakistan along ethnic lines as a result of "ethnic differences" must be viewed in the wider context of geopolitical ambitions. His article lays the foundation both for the explanation of a potential breakup, and thus the "justification" for Western intervention in the conflict. His "predictive" capacities as a seasoned journalist can be alternatively viewed as pre-emptive imperial propaganda.

FRACTURING PAKISTAN

The war in Afghanistan is inherently related to the situation in Pakistan. From the days of the Afghan-Soviet war in the 1980s, arms and money were flowing through Pakistan to the Mujahideen in Afghanistan. During the civil war that followed, Pakistan armed and financed the Taliban, which eventually took power. When the U.S. and NATO initially attacked Afghanistan on October 7, 2001, this was primarily achieved through cooperation with Pakistan. When the war theatre was re-named "AfPak," the role of Pakistan, however, was formally altered. While the previous few years had seen the implementation of a strategy of destabilizing Pakistan, once the "AfPak" war theatre was established, Pakistan ceased to be as much of a conduit or proxy state and became a target.

In September of 2008, the editor of Indian Defence Review wrote an article explaining that a stable Pakistan is not in India's interests: "With Pakistan on the brink of collapse due to massive internal as well as international contradictions, it is matter of time before it ceases to exist." He explained that Pakistan's collapse would bring "multiple benefits" to India, including preventing China from gaining a major port in the Indian Ocean, which is in the mutual interest of the United States. The author explained that this would be a "severe jolt" to China's expansionist aims, and further, "India's access to Central Asian energy routes will open up."

In August of 2009, Foreign Policy Journal published a report of an exclusive interview they held with former Pakistani ISI chief Lieutenant General Hamid Gul, who was Director General of the powerful intelligence services (ISI) between 1987 and 1989, at a time in which it was working closely with the CIA to fund and arm the Mujahideen. Once a close ally of the US, he is now considered extremely controversial and the US even recommended the UN to put him on the international terrorist list. Gul explained that he felt that the American people have not been told the truth about 9/11, and that the 9/11 Commission was a "cover up," pointing out that, "They [the American government] haven't even proved the case that 9/11 was done by Osama bin Laden and al Qaeda." He said that the real reasons for the war on Afghanistan were that:

the U.S. wanted to "reach out to the Central Asian oilfields" and "open the door there", which "was a requirement of corporate America, because the Taliban had not complied with their desire to allow an oil and gas pipeline to pass through Afghanistan. UNOCAL is a case in point. They wanted to keep the Chinese out. They wanted to give a wider security shield to the state of Israel, and they wanted to include this region into that shield. And that's why they were talking at that time very hotly about 'greater Middle East'. They were redrawing the map."

He also stated that part of the reason for going into Afghanistan was "to go for Pakistan's nuclear capability," as the U.S. "signed this strategic deal with India, and this was brokered by Israel. So there is a nexus now between Washington, Tel Aviv, and New Delhi." When he was asked about the Pakistani Taliban, which the Pakistani government was being pressured to fight, and where the financing for that group came from; Gul stated:

Yeah, of course they are getting it from across the Durand line, from Afghanistan. And the Mossad is sitting there, RAW is sitting there — the Indian intelligence agency — they have the umbrella of the U.S. And now they have created another organization which is called RAMA. It may be news to you that very soon this intelligence agency — of course, they have decided to keep it covert — but it is Research and Analysis Milli Afghanistan. That's the name. The Indians have helped create this organization, and its job is mainly to destabilize Pakistan.

He explained that the Chief of Staff of the Afghan Army had told him that he had gone to India to offer the Indians five bases in Afghanistan, three of which are along the Pakistani border. Gul was asked a question as to why, if the West was supporting the TTP (Pakistani Taliban), would a CIA drone have killed the leader of the TTP. Gul explained that while Pakistan was fighting directly against the TTP leader, Baitullah Mehsud, the Pakistani government would provide the Americans where Mehsud was, "three times the Pakistan intelligence tipped off America, but they did not attack him." So why all of a sudden did they attack?

Because there were some secret talks going on between Baitullah Mehsud and the Pakistani military establishment. They wanted to reach a peace agreement, and if you recall there is a long history of our tribal areas, whenever a tribal militant has reached a peace agreement with the government of Pakistan, Americans have without any hesitation struck that target.

... there was some kind of a deal which was about to be arrived at – they may have already cut a deal. I don't know. I don't have enough information on that. But this is my hunch, that Baitullah was killed because now he was trying to reach an agreement with the Pakistan army. And that's why there were no suicide attacks inside Pakistan for the past six or seven months.

An article in one of Canada's national magazines, Macleans, reported on an interview with a Pakistani ISI spy, who claimed that India's intelligence services, Research and Analysis Wing (RAW), have "tens of thousands of RAW agents in Pakistan." Many officials inside Pakistan were convinced that, "India's endgame is nothing less than the breakup of Pakistan. And the RAW is no novice in that area. In the 1960s, it was actively involved in supporting separatists in Bangladesh, at the time East Pakistan. The eventual victory of Bangladeshi nationalism in 1971 was in large part credited to the support the RAW gave the secessionists."

Further, there were Indian consulates set up in Kandahar, the area of Afghanistan where Canadian troops are located, and which is strategically located next to the Pakistani province of Baluchistan, which is home to a virulent separatist movement, of which Pakistan claims is being supported by India. Macleans reported on the conclusions by Michel Chossudovsky, economics professor at University of Ottawa, that, "the region's massive gas and oil reserves are of strategic interest to the U.S. and India. A gas pipeline slated to be built from Iran to India, two countries that already enjoy close ties, would run through Baluchistan. The Baluch separatist movement, which is also active in Iran, offers an ideal proxy for both the U.S. and India to ensure their interests are met."

Even an Afghan government adviser told the media that India was using Afghan territory to destabilize Pakistan. In September of 2009, the Pakistan Daily reported that captured members and leaders of the Pakistani Taliban have admitted to being trained and armed by India through RAW or RAMA in Afghanistan in order to fight the Pakistani Army.

Foreign Policy magazine in February of 2009 quoted a former intelligence official as saying, "The Indians are up to their necks in supporting the Taliban against the Pakistani government in Afghanistan and Pakistan," and that, "the same anti-Pakistani forces in Afghanistan also shooting at American soldiers are getting support from India. India should close its diplomatic establishments in Afghanistan and get the Christ out of there."

The Council on Foreign Relations published a backgrounder report on RAW, India's intelligence agency, founded in 1968 "primarily to counter China's influence, [however] over time it has shifted its focus to India's other traditional rival, Pakistan." For over three decades both Indian and Pakistani intelligence agencies have been involved in covert operations against one another. One of RAW's main successes was its covert operations in East Pakistan, now known as Bangladesh, which "aimed at fomenting independence sentiment" and ultimately led to the separation of Bangladesh by directly funding, arming and training the Pakistani separatists. Further, as the Council on Foreign Relations noted, "From the early days, RAW had a secret liaison relationship with the Mossad, Israel's external intelligence agency."

Since RAW was founded in 1968, it had developed close ties with the Afghan intelligence agency, KHAD, primarily to do with intelligence sharing on Pakistan. In the 1980s, while Pakistan was funding, arming and training the Afghan Mujahideen with the support of Saudi Arabia and the CIA, India was funding two covert groups which orchestrated terrorist attacks inside Pakistan, which included a "low-grade but steady campaign of bombings in major Pakistani cities, notably Karachi and Lahore." RAW has also had a close relationship with the CIA, as even six years before RAW was created, in 1962, the CIA created a covert organization made up of Tibetan refugees, which aimed to "execute deep-penetration terror operations in China." The CIA subsequently played a part in the creation of RAW. In the 1980s, while the CIA was working closely with the ISI in Pakistan, RAW, while wary of their relationship, continued to get counterterrorism training from the CIA.

In October of 2009, the New York Times reported that the US strategy "to vastly expand its aid to Pakistan, as well as the footprint of its embassy and private security contractors here, are aggravating an already volatile anti-American mood as Washington pushes for greater action by the government against the Taliban." The U.S. gave Pakistan an aid deal of $1.5 billion per year for the next five years, under the stipulation of "Pakistan to cease supporting terrorist groups on its soil and to ensure that the military does not interfere with civilian politics." President Zaradari accepted the proposal, making him even more unpopular in Pakistan, and further angering Pakistan's powerful military, which sees the deal as interfering in the internal affairs of the country.

America is thus expanding its embassy and security presence within the country, as the Embassy "has publicized plans for a vast new building in Islamabad for about 1,000 people, with security for some diplomats provided through a Washington-based private contracting company, DynCorp." The NYT article referred to how relations were becoming increasingly strained between Pakistan and the US, and tensions were growing within the country exponentially, as "the American presence was fuelling a sense of occupation among Pakistani politicians and security officials," and several Pakistani officials stated that, "the United States was now seen as behaving in Pakistan much as it did in Iraq and Afghanistan." Futher:

In particular, the Pakistani military and the intelligence agencies are concerned that DynCorp is being used by Washington to develop a parallel network of security and intelligence personnel within Pakistan, officials and politicians close to the army said.

The concerns are serious enough that last month a local company hired by DynCorp to provide Pakistani men to be trained as security guards for American diplomats was raided by the Islamabad police. The owner of the company, the Inter-Risk Security

Company, Capt. Syed Ali Ja Zaidi, was later arrested. The action against Inter-Risk, apparently intended to cripple the DynCorp program, was taken on orders from the senior levels of the Pakistani government, said an official familiar with the raid, who was not authorized to speak on the record.

The entire workings of DynCorp within Pakistan are now under review by the Pakistani government.

As revealed in the Wikileaks diplomatic cables, U.S. Ambassador to Pakistan Anne Patterson wrote in September of 2009 that the U.S. strategy of unilateral strikes inside Pakistan "risk destabilizing the Pakistani state, alienating both the civilian government and military leadership, and provoking a broader governance crisis in Pakistan without finally achieving the goal."

In an interview with Press TV, Hamid Gul, former Inter-Services Intelligence chief revealed more of what he sees as the US strategy in Pakistan. He explained that with the massive expansion of the U.S. Embassy in Pakistan, and alongside that, the increased security staff, the Chinese are becoming increasingly concerned with the sovereignty and security of Pakistan. He claimed that the money that the US government offered (with heavy conditions) to Pakistan, $1.5 billion every year for five years, will be spent under the direction of the Americans, and that "they are going to set up a large intelligence network inside Pakistan," and ultimately "they really want to go for Pakistan's nuclear assets." He further claimed that the Indians are trying to destabilize Pakistan; however, he explained, this does not necessarily mean disintegrate, but rather: they are trying to destabilize Pakistan at the moment so that it feels weak and economically has to go begging on its knees to Americans and ask for succor and help. And in that process they will want to expect certain concessions with regards to nuclear power and also with regards to setting up their facilities here in Pakistan.

When he was asked what America's long-term goal was in regards to Pakistan, Gul responded that the goal: for America is that they want to keep Pakistan destabilized; perhaps create a way for Baluchistan as a separate state and then create problems for Iran so that this new state will talk about greater Baluchistan... So it appears that the long-term objectives are really to fragment all these countries to an extent that they can establish a strip that would be pro-America, pro-India, pro-Israel. So this seems to be their long-term objective apart from denuclearizing Pakistan and blocking Iran's progress in the nuclear field.

BATTLE IN INDIA OVER BALOCHISTAN

Hell has no fury like an offended far-right nationalist. And even a democracy can deny one the power to dissent when ltra-nationalists are out to exploit an issue involving a country's traditional "enemy." India provides a current illustration. For days now, the country's parliament has been rocked by a controversy over a joint statement of the prime ministers of India and Pakistan. As noted in these columns before (A Piece of Paper and South Asian Peace, July 20, 2009), the enemies of South Asian peace began by attacking the document for "delinking" the issue of terrorism from the India-Pakistan "composite dialogue" suspended since the Mumbai outrage of eight months ago.

The focus of the far-right fury has now shifted to the last lines of the statement lambasted as a "sell-out." The brief document is denounced as all the more of a "betrayal" for mentioning Pakistan's province of Balochistan, with a hint at India's alleged hand in the Baloch insurgency.

The statement, issued on July 16 after a meeting between Manmohan Singh and Yousaf Raza Gilani at the non-aligned summit in Sharm el-Sheikh (Egypt), ends with the words: "Prime Minister Gilani mentioned that Pakistan has some information on threats in Balochistan and other areas." Working itself up into a white heat of rage over this, the main opposition Bharatiya Janata Party (BJP) and its allies have staged a series of walkouts in the two houses of the parliament, with its media friends trying to whip up a furious campaign on the issue.

"This has never happened before" was their near-hysterical outcry. This is not the first time Pakistan has complained about India's involvement in its south-western province. From 2004, in fact, both local and national leaders, including former President Pervez Musharraf, have leveled the allegation in an increasingly strident manner. India has been charged not only with subversive operations from within Balochistan, but also with running camps for training Baloch insurgents in neighbouring Afghanistan.

It is true, however, that Balochistan has never figured in an India-Pakistan joint statement before, Nor indeed has Islamabad taken its complaint on this count to any international forum, including the United Nations. But Pakistan may not have gained from inclusion of the issue in the joint statement, as we shall see just in a while. Meanwhile, the BJP seemed to receive fresh ammunition from reports in the Pakistani media that Gilani confronted Singh with a dossier detailing India's covert subversion in the restive province. According to the reports, the dossier listed the safe houses being run by India's external intelligence agency, the Research and Analysis Wing (RAW) of its External Affairs Ministry, in Afghanistan for training Baloch militants for "missions in Pakistan."

According to one report, Pakistani Foreign Secretary Salman Bashir told Indians in Egypt that, if India did not agree to the "delinking," Pakistan would be forced to produce before the international media at least "three Indian Ajmal Kasabs" who were directly or indirectly part of the terrorist activities in Balochistan. This alluded to Ajmal Kasab, a Pakistani who is the only militant involved in Mumbai caught alive and is undergoing a trial now in India.

The alleged dossier has slowly disappeared from the debate. First, it was claimed that the dossier was only shown, and not given, to Singh. Now it has been denied by both sides. What happened in the meeting, then, to warrant those lines? Singh's explanation in the parliament on July 30 was: "When I spoke to Prime Minister Gilani about terrorism from Pakistan, he mentioned to me that many Pakistanis thought that India meddled in Balochistan. I told him that we have no interest in destabilizing Pakistan.... If Pakistan has any evidence,... we are willing to look at it because we have nothing to hide."

Gilani - whatever else might be his game - was not wrong about the perception on the issue in Pakistan. Indian peace activist and Magsaysay award-winner Sandeep Pandey, just back from a visit to Pakistan, says: "Manmohan Singh has merely acknowledged something which is common knowledge in Pakistan." It is not something unknown to concerned Indian agencies and quarters close to them, either.

It would be terribly unfair, of course, to tar all Balochs fighting for independence or autonomy as terrorists. Balochistan is the largest as well as the least-developed province of Pakistan. It is certainly not without reason that insurgency has continued here relentlessly since the late 1940's. All this, however, is no argument for rejecting evidence of at least India's past role in Balochistan, available from Indian sources of proximity to undeclared policies. A journalist known for closeness to security and intelligence agencies, Praveen Swami, wrote a year ago: "As the Nawab Akbar Khan Bugti-led Baloch insurgency against Pakistan escalated from 2004, Islamabad's allegations grew more strident. In August 2004, Balochistan Chief Minister Jam Mohammad Yusuf declared that India was running 40 terror camps targeting the province.... At the beginning of July 2006, Mushahid Husain, the chair of the Pakistan Senate Standing Committee on Foreign Affairs, asserted that RAW was 'training 600 Balochis' in Afghanistan."

Significantly, Swami adds: "Were these charges true? It is hard to say. Credible allegations exist that India offered low-grade funding for Baloch insurgents in the wake of the 1971 Bangladesh war of liberation - but then withdrew support to avoid destabilizing Prime Minister (Zulfikar Ali) Bhutto. It is probable, though not proved, that India also held out some financial support to Bugti and other Baloch nationalist leaders."

In February 2006, the well-informed Swami had something more shocking to say: "Musharraf has not made public the evidence he claims to possess on Indian involvement in Balochistan, and leaders of the insurgent groups there have been dismissive of his assertion.... Yet, it is also true that Indian frustration with Musharraf's refusal to deliver on his promises to end Islamist terrorism in Jammu and Kashmir has been growing. Many in India's intelligence establishment believe that the time has come to use what are called offensive covert means."

Explains Swami: "In the past, India has demonstrated the capability to do so, and with good effect: two Research and Analysis

Wing (RAW) groups, CIT-'X' and CIT-'J', had, for example, carried out a series of bombings in Lahore and Karachi as retaliation for strikes by Pakistan-backed Khalistan terror groups. India's offensive covert operations were closed down early in the tenure of Prime Minister I. K. Gujral, but the weapon, as Musharraf well knows, could be unsheathed again." This could be done, he meant, in Balochistan.

Security analyst Ravi Visvesvarya Prasad voices a similar grievance about Gujral, vindicating critics of India's past activities in Balochistan. In a newspaper article of July 27, 2009, he writes: "... India's covert capabilities in Pakistan, painstakingly built up during the regimes of (former Prime Ministers) Indira Gandhi and P. V. Narasimha Rao, were unilaterally shut down by then-Prime Minister I. K. Gujral in 1997-98 as part of his peace initiative. As a result, India today has virtually no reliable and capable human intelligence assets within Pakistani territory."

A related point, with an added religious dimension, is made by former RAW operative and security expert B. Raman in a recent analysis. Says he: "The Hindu population of Balochistan has considerably come down in recent years due to their forcible ejection from there by the Army after it launched the project for the construction of the Gwadar port with Chinese assistance. The Inter-Services Intelligence looked upon them as potential Research and Analysis Wing agents and forced them to shift."

Raman adds: "But the Hindus who have managed to remain have continued to enjoy the protection of the Baloch leaders and people. When Nawab Akbar Khan Bugti was killed by the Pervez Musharraf government in a military operation in August 2006, among those who died with him were Baloch Hindus whom he had taken under his protection."

Raman outdoes the BJP in that he considers the joint statement a betrayal of Baloch "freedom fighters" as well, since it equates them with "terrorists" and thus serves the interests of Islamabad. But, as mentioned before, Balochs themselves do not seem to share his bitterness.

An organization of the indigenous Marri people of the province, Anjuman Ittehad Marri (AIM), says: "The mention of Baluchistan in the joint statement... has brought the goal of an independent Baluchistan even closer as it has helped internationalize the Baluch question..."

Hamid Mir of Pakistan's increasingly influential Geo TV agrees with the AIM assessment, while explaining why Pakistani rulers have been reluctant to expose India's role in Balochistan. Says he: "... Balochistan is not a serious dispute like Jammu and Kashmir; it is a problem of provincial rights. Instead of internationalizing the issue, therefore, Islamabad should address the problem realistically. It cannot get away by blaming India alone for the unrest in the province."

It would serve better the cause of both the countries - and South Asia - if they go for bilateral talks on both issues instead of internationalizing the disputes further. They cannot succeed, however, if they seek to do so behind the backs of the people of either Kashmir or Balochistan.

ROLE OF RAW IN BALOCHISTAN

Against great odds, the Baloch freedom-fighters, who started their second war of independence in 2005, have managed to keep up their valiant struggle for the independence of their homeland.

Despite repression and torture by the Pakistani Army, its Frontier Corps and the Inter-Services Intelligence (ISI), the freedom-fighters, operating in different groups under different names, have managed to steadily weaken the hold of the Pakistani security forces and intelligence in their homeland and challenge the attempts of the Pakistani authorities to deprive the Balochs of their naturalresources such as the Sui gas which were being diverted for the benefit of the Punjabis.

Despite repeated expressions of their interest in finding a political solution to the grievances of the Balochs ranging from greater and genuine autonomy to independence, the mainstream political parties of Pakistan such as the Pakistan People’s Party of President Asif Ali Zardari and the two factions of the Pakistan Muslim League (PML) led by Mr. Nawaz Sharif and Chaudhury Shujat Hussain have not taken any action to withdraw the security forces from Balochistan, release all political prisoners and put an end to the obnoxious practice of missing persons. Under this practice, Baloch youth perceived as hostile to the Army and the ISI are kidnapped by the intelligence agencies for illegal detention. They are projected by the police as having joined the militancy and hence missing from their homes. Weeks and months later, their mutilated bodies are found thrown in remote areas of the province.

The Pakistani Supreme Court as well as Pakistani human rights organisations and activists have increasingly drawn attention to the serious violations of the human rights of the Balochs and to the continuing obnoxious practice of missing persons. The Baloch grievances now receive greater attention from enlightened sections of the Pakistani society as well as political elements and human rights activists in the US and the European Union countries. In recent months, there was greater sympathy for Baloch rights in some sections of the US Congress.

The Baloch freedom-fighters operating from Balochistan and Baloch political activists in the Baloch diaspora in the US and Europe have for the first time succeeded in creating a growing understanding of the Baloch cause and the Baloch political and economic grievances.

The deterioration in the relations between Washington DC and Islamabad after the Abbottabad raid of May 2 last year by the US Navy Seals to kill Osama bin Laden also created for the first time greater empathy for the Baloch cause in sections of the US political circles. There was greater interest in what was happening in Balochistan and a greater dissemination of correct information regarding the real state of affairs in Balochistan. This was a positive gain for the Baloch freedom-struggle and a significant strategic step forward. From a war of tactics, the freedom-struggle seemed tobe moving towards a strategic war with greater thinking and debate on a Baloch vision of the future.

No country in the world was officially prepared to support the Baloch freedom struggle. No State could have supported the freedom struggle and the objective of an independent Balochistan. That could have been projected by Islamabad as amounting to interference in the internal affairs of Pakistan. But there was growing sympathy for the Baloch cause and vision in sections of the civil societies of different countries.

Despite these strategic gains of recent months, the Baloch freedom-fighters are still far from their ultimate objective of an independent Balochistan. Lack of unity among different organisations in Balochistan itself and among different diaspora groups abroad continue to stand in the way of unity of action to give a final decisive push forward to the independence struggle. Unless the persisting factionalism and ego clashes among different Baloch leaders is put an end to and a unified leadership starts guiding the movement, the enormous sacrifices being made by Balochistan;s GenNext may prove to be in vain. This is an issue that needs attention from all Baloch leaders and activists.

They already seem to have missed the window of opportunity provided by the intense distrust of the Pakistani Army and Intelligence in the US political and administrative circles after the Abbottabad raid and the US discovery that the Pakistan Army and its ISI, while pretending to fight against Al Qaeda and its associates, have actually been acting in collusion with them.

As the US Presidential elections approach in November next, the US itself as well as other NATO countries seem to be in a hurry to start the process of disengagement from Afghanistan. The recent decision by the US to mend fences with Pakistan at least tactically if not strategically to facilitate the on-coming process of disengagement is likely to have unfortunate consequences for the Baloch freedom struggle. Those in the US, who had started evincing increasing interest in the Baloch cause and struggle, may start developing second thoughts about the wisdom of continuing to show empathy for the Balochs at a time when the US might need the co-operation of the Pakistani State to disengage from Afghanistan. At this time, when greater Baloch unity is needed, if the different groups continue to indulge in ego clashes, the State of Pakistan might be the beneficiary.

Since the beginning of this year, the Pakistani Army and ISI have been following an insidious new policy in Balochistan in an attempt to discredit the Baloch freedom-fighters in the eyes of Western public opinion as terrorists. There has been a wave of massacres of Shias in Balochistan—some settlers from Punjab and many members of the Hazara tribal community. Over 80 Shia fatalities have been reported in the last six months as a result of the targeted attacks on the Shias.

The Sunni extremist Lashkar-e-Jhangvi (LEJ) of Pakistan, which had helped the Afghan Taliban in carrying out massacres of the Hazara Shias of Afghanistan before the Taliban was overthrown in 2001, has claimed responsibility for some of these massacres in Balochistan. The Pakistan Army and ISI have not taken any action against the LEJ and to protect the Shias. Instead, one could discernthe beginning of an insidious campaign by the ISI to project these massacres as carried out by sections of the Baloch freedom fighters because many of the Shias had migrated from Punjab. This insidious campaign needs to be effectively countered by the Baloch activists.

As the NATO powers start the process of disengaging from Afghanistan and seek the co-operation of the Pakistan Army and ISI for this purpose, support for the Baloch cause in sections of Western societies may freeze and even start declining. How to prevent this is a question that needs the serious attention of the strategists of the Baloch freedom struggle. The Baloch vision will remain the same, but there is going to be a need for a new Baloch strategy to give shape to this vision at a time of unpredictable transition in Afghanistan.

10

Bangladesh: ISI, RAW, DGFI, ULFA and MJC

For the first time, several world leaders, including United Nations Secretary General Kofi Annan and British Foreign Minister Jack Straw, and the United States have publicly focused their attention on India's turbulent northeast and its insurgent violence. "It is really a matter of serious concern that several anti-India forces of China, Myanmar, Pakistan and Bangladesh are trying hard to create unrest in northeastern states by using the fundamentalists and separatist forces of the region." Al-Qaeda and its associates now have a sizable presence in Bangladesh. It serves as a focal point or umbrella organization for a worldwide network that includes many Islamic extremist groups such as Harkat-ul-Mujahideen, which has bases in India's northeastern states also. The ULFA's identification with the known Pakistani ambition of creating a 'Swadhin Asom' [Independent Sovereign Assam] that would be a full-fledged Muslim-majority state through a demographic invasion and Bangladeshi dreams of a Brihot [Greater] Bangladesh is now so complete that it has even changed its original aim of fighting and expelling Muslim infiltrators and illegal IMMIGRANTS from Bangladesh to giving them membership of the organization."

New Delhi and Washington are finally waking up to the strange phenomenon taking place in India's troubled northeast. An upper-caste Hindu militant secessionist organization of the northeastern state of Assam - the United Liberation Front of Asom (ULFA) -

has reportedly taken up formal membership with the Pakistan-based Muttahida (United) Jihad Council (MJC) after years of dallying with the Pakistani military's Inter-Services Intelligence (ISI) and the Directorate General of Field Intelligence (DGFI) of Bangladesh.

The MJC is an umbrella organization of various outfits engaged in militant operations in the Indian state of Jammu and Kashmir. Its most prominent member group is the dreaded Hizbul Mujahideen, headed by Syed Salahuddin.

For the first time, several world leaders, including United Nations Secretary General Kofi Annan and British Foreign Minister Jack Straw, and the United States have publicly focused their attention on India's turbulent northeast and its insurgent violence. US Ambassador to India David C Mulford offered the help of Federal Bureau of Investigation (FBI) members after the serial blasts on October 2 that killed scores of people. In a letter to Assam Chief Minister Tarun Gogoi, Mulford said, "Should you find it helpful, the FBI would be pleased to provide technical support for your investigation."

Though Gogoi welcomed the offer, it was criticized by leftist as well as Hindu-right politicians and sections of the media as interfering in India's internal matters, an infringement of sovereignty and at the very least "unmindful of the established procedures for diplomatic conduct in the host country", which clearly specify that all communication between an envoy and a state government must be routed through the Ministry of External Affairs. However, a realization is now dawning that this offer is probably an indication of the United States' recognition of a global and Islamist dimension to the terrorism in India's northeast.

Indicating that the government of India is not only aware of the foreign dimension of the terror-bombings in Assam and Nagaland, but it is also willing to take some concrete action, the normally reticent Indian Defense Minister Pranab Mukherjee put the blame for these terrorist acts squarely on the ISI of Pakistan. Indian Prime Minister Manmohan Singh, too, has now voiced his concern over "some insurgent groups" taking "shelter" in Bangladesh. The chief ministers of Assam and Tripura have been demanding that Bangladesh be persuaded to take the kind of robust action against the insurgents based on its territory that Bhutan took last year. Indeed, part of the reason for this latest concentration of militants in Bangladesh and Myanmar is that they lost their sanctuaries in Bhutan.

Assam Chief Minister Tarun Gogoi has linked the attacks to the ULFA regrouping after being flushed out of Bhutan last year. He said: "The Bhutanese military offensive against ULFA and NDFB [National Democratic Front of Bodoland] inside the kingdom in December last year has not helped us. The rebels appear to have simply relocated their camps." He is pushing New Delhi to focus on Bangladesh and Myanmar, where he insists ULFA and NDFB have bases and safe houses. "As we see now, the root of the problem lies in these two countries. Unless the rebels are uprooted from there, violent attacks are bound to go on," Gogoi said.

Director general of Tripura police G M Srivastava, however, implicated Chinese intelligence, too, in the growing militancy in the region. He said: "It is really a matter of serious concern that several anti-India forces of China, Myanmar, Pakistan and Bangladesh are trying hard to create unrest in northeastern states by using the fundamentalists and separatist forces of the region."

Stressing the need for strong diplomatic pressure on Dhaka to contain these forces, he said: "We have got the photographs of training camps in Dhaka where cadres of Indian insurgent groups are also being trained by operatives of the Taliban and the al-Qaeda network."

Al-Qaeda and its associates now have a sizable presence in Bangladesh. It serves as a focal point or umbrella organization for a worldwide network that includes many Islamic extremist groups such as Harkat-ul-Mujahideen, which has bases in India's northeastern states also. Concerning the operatives from China, Srivastava said: "Their presence is in the field but they are very discreet kind of operatives so it is very difficult to get direct evidence against China but there are a number of circumstantial evidences which indicate their involvement."

The chief minister's view is seconded by Khagen Sharma, inspector general of the Assam police intelligence wing. "In

Bangladesh, the ULFA is in the care of Pakistan's ISI. In Myanmar, across Arunachal Pradesh, the ULFA is being provided logistical support by the Khaplang faction of the National Socialist Council of Nagaland," said Sharma.

Bangladesh, however, has consistently denied the presence of any such sanctuaries and training camps. Indeed, senior Bangladeshi journalist Enayetullah Khan claimed that India's external intelligence arm, the Research and Analysis Wing (R&AW), has much deeper penetration, network and wherewithal in Bangladesh to "beat the ISI hollow as far as moles and operatives and also clandestine training are concerned".

New Delhi has on several occasions furnished detailed lists of rebel camps inside Bangladesh to authorities there. The last such list of 195 camps was handed over to visiting Bangladesh Rifles chief Major-General Jehangir Alam Choudhuri by his Border Security Force counterpart Ajai Raj Sharma in New Delhi last month. According to this list, two militant groups of the Indian state of Tripura alone are running training camps in the Bangladeshi districts of Moulvi Bazar, Habibganj, Comilla, Rangamati, Khagracherry, Chittagong and Cox's Bazar. The list, however, merely invited a derisive rejection from Dhaka. On checking the provided coordinates, they said one training camp on the list would have to be within a Bangladesh Rifles cantonment and another at sea.

New Delhi has asked the Intelligence Bureau (IB), India's domestic intelligence agency, to confirm the recent move of ULFA to take up the membership of the Jihad Council with "concrete evidence". It has also asked the IB to carry out a "special survey" of the entire northeastern region, focusing on detailed information on insurgent groups, their hierarchy, location of training camps. The security agencies had recently reported to the Ministry of Home Affairs that the ULFA had accepted membership of the MJC for logistics support.

ULFA is the biggest among scores of terrorist organizations operating in the troubled northeast of India that borders Bangladesh, China, Myanmar, Bhutan and Nepal. Though the Indian security agencies have long known about ULFA's links with

Pakistani and Bangladeshi intelligence, and have been warning the government for at least 18 months, they received word of ULFA's jihadi links in September after certain seizures made from the hideouts of a slain ULFA militant. A grenade found from militant Lachit Rava, who was killed in an encounter at Jerdoba in East Garo Hill district in Meghalaya, India, on September 14, was similar to those used by militants during the attack on the Indian parliament in December 2002. The Pakistani army is also known to use such explosives. The discovery that northeastern militants are now using programmable time-delay devices made in Pakistan is considered particularly significant. The seizures made during subsequent raids further proved the banned outfit's close links with Pakistan-based jihadi groups.

Although the IB has already confirmed these reports, a senior Home Ministry official told Asia Times Online the government now wants the information to be backed by evidence so that the matter can be taken up at the diplomatic level. In view of similar reports about other organizations, the government wants to deal with the problem more comprehensively and has ordered the preparation of a complete profile of about 108 banned groups in the region.

Terrorist attacks in different parts of India's northeast, including serial bomb blasts and gun attacks in Assam and Nagaland, killed more than 80 people and wounded 200 in a span of just 72 hours beginning on October 2, the day India remembers the apostle of non-violence, the father of the nation, Mahatma Gandhi.

The outlawed ULFA, fighting for a "sovereign, socialist Assam" since its inception in 1979, and the NDFB, pushing an armed campaign for a separate Bodo homeland since 1986, have claimed responsibility for some of the attacks.

According to a Home Ministry official who has dealt with the northeast for several years, the government has been aware for quite some time of the ULFA and other similar organization's links to ISI and DGFI. The government analysts can even understand the necessity of such links from the secessionists' point of view; their need for arms and ammunition and other logistical and even support: but reports of their joining the Jihad Council has completely mystified them. The official told Asia Times Online: "The ULFA's identification with the known Pakistani ambition of creating a 'Swadhin Asom' [Independent Sovereign Assam] that would be a full-fledged Muslim-majority state through a demographic invasion and Bangladeshi dreams of a Brihot [Greater] Bangladesh is now so complete that it has even changed its original aim of fighting and expelling Muslim infiltrators and illegal from Bangladesh to giving them membership of the organization."

No one seems to understand what an Assamese movement of upper-caste Hindus, even if militant or secessionist, some of whose leaders were educated in reputed Indian educational institutions, would gain by helping create a new and sovereign Muslim state or a Greater Bangladesh.

The ULFA has indeed changed not only its aims and objects formally but also the very definition of the word "Asamiya" (Assamese) in order to accommodate illegal Muslim from Bangladesh. (Incidentally, the influx of Hindu Bangladeshis was never a problem as they were considered refugees fleeing discrimination and persecution in Muslim-majority Bangladesh.)

In the latest ULFA membership form the very first clause seeks to accommodate all nationalities. It says that while a member "must be a permanent resident of Assam, volunteers from other countries will be recognized conditionally". Similarly, the concept of Assamese is redefined by stating that "the Asamiya are a people of all communities, the mixture of people who are determined to work for all-round progress of Assam". Thus the concept of Assamese is no longer restricted, as earlier, to those speaking the Assamese language as their mother tongue. It obviously includes the Bangla-speaking people as well. This is further clarified a little later in a paragraph beginning with the sentence: "The contribution of the people of East Bengal [Bangladesh] origin in Assam towards increasing the state's economic output is indeed noteworthy."

Contrasting this with the ULFA's original "aims and objectives" will leave no one in doubt that it is now completely under the sway of its Pakistani and Bangladeshi masters. Its original manifesto justified its founding in 1979 as a fight against the "influx of foreigners and massive exploitation of its natural resources".

Seeking to justify its armed struggle, the ULFA manifesto says: "Especially in 1979, democratic and unarmed peaceful movement against the influx of foreigners and economic exploitation, the occupation force of India killed 700 unarmed agitators where the majority were teenage students. Though the people of Assam and leadership of the struggle have a strong stand for peaceful and amicable solution of the conflict, India has always been trying to force a military solution, thus the unarmed peaceful movement against the influx and economic exploitation transformed to an armed national liberation struggle."

It was because of this strong orientation of the ULFA against the influx of Muslim Bangladeshis that the entire Assam students' movement of 1979 and early 1980s, of which it was a product, was branded "Hindu fundamentalist" despite its original secular track. Interestingly, it now says: "We would like to state here for everybody's information that the movement led by the All Assam Students Union and Gana Sangram Parishad from 1979 to 1985 is viewed by the ULFA as one based on emotion." It now denounces the Assam movement as "one that was led by juveniles, who failed to understand that migration per se was not bad and had helped many countries like the USA to become what they are today". Nothing could mark the abdication of ULFA's original platform and history as completely as this one sentence. The ULFA is thus no longer silent on the issue of illegal Muslim, as it was during the late 1980s and early 1990s, and is now making its intentions clear.

Having read these proclamations, it becomes possible for government officials to believe, though still not understand, reports that the ULFA is actually running schools in Dhaka training poor Bangladeshis to infiltrate Assam. The course they teach includes imparting some familiarity with the spoken Assamese language and a smattering of the written language, at least for them to be able to read road signs and names of shops, etc.

Another significant indication of the ISI's grip over ULFA comes from the fact that unlike in the past, it is no longer bothered about killing innocents. Regarding itself a freedom movement it used to avoid killing as much as possible for fear of backlash from the common people. But it is clear from the August 15 blast in Dhemaji - killing scores of schoolchildren - and the same message underscored by subsequent blasts, that it no longer has such scruples. This is very much in tune with ISI strategy, as seen in Kashmir, of creating maximum chaos and bleeding India through "a thousand cuts".

Meanwhile, New Delhi on Friday said it is willing to join three-way talks with Assam and separatist rebels to end violence in the state. The NDFB agreed to talks this month.

THE FREEDOM LABORATORY

At last, more than four decades after 1971, we have the first important international histories of the Bangladesh liberation war based on archival materials. Srinath Raghavan has delved widely into the archives of several countries to produce a 'global history'. Gary Bass focuses on US and Indian policies. Both have made full use of the Indian records available in the Nehru Memorial Museum and Library as well as the recently declassified ministry of external affairs files housed in the National Archives.

Raghavan has produced a scholarly study couched in sparkling prose. He has a wide canvas, seeking not only to delineate the policies of the major international actors, including a number of middle powers, but also to situate the liberation struggle in the context of broader global historical processes. He is at his best as a diplomatic historian. The centrepiece of his book is a detailed, skilfully pieced together account of the evolution of Indian, US and Russian policies in 1971. He also captures in perfect cameos—brief but precise—the policies of UK, France, frg, gdr and Australia. He is less successful in grasping the deep historical roots of the Bangladesh movement. Partly as a consequence, his ambitious attempt to locate the liberation war within the context of global trends is not always convincing.

The narrative begins somewhat abruptly in 1969 with Ayub's downfall, an event which Raghavan believes "was the critical turning point that led toward the road to Bangladesh". In reality, the foundations of united Pakistan had become very shaky by then, leading RAW as well as US State Department analysts to question its survival. In February 1969, the director of intelligence and research in the State Department was already reporting that "secession cannot be ruled out". In Raghavan's account, by January 1971, "RAW had begun to receive inputs suggesting that Mujib considered secession to be a real possibility" [emphasis added]. In fact, such inputs were being received long before 1971. In 1966, Sheikh Mujibur Rahman told deputy high commissioner Asoke Ray that the six-point programme was intended as a bridge to independence.

However, coming to the events of 1971, Raghavan weaves together a compelling account of the evolution of Indian policy in a rapidly changing scene. He explains why, before the brutal military crackdown on March 25, the "possibility of secession by East Pakistan was recognised—only to be assessed as antithetical to India's interests". He punctures Manekshaw's boast that he dissuaded Indira Gandhi from prematurely launching a military intervention. There is an excellent account of the origins of the Indo-Soviet Treaty and India's diplomatic and military preparations to assist the liberation struggle. Raghavan provides conclusive evidence that India never had West Pakistan in its sights and that the CIA report to the contrary was fictitious.

There are some important omissions in Raghavan's account. With the approach of the *USS Enterprise*, India requested Moscow to make a public announcement at the highest level, warning against involvement or interference by third countries. Raghavan states that Moscow "declined to make the announcement". But he misses the fact that the Soviet leadership promptly sent a top-secret message to Nixon on these lines. They also ostentatiously enhanced their naval presence in the Indian Ocean. After his discussions with Zhou Enlai in July 1971, Kissinger came away with the impression that China would intervene in the event of an Indo-Pakistan war. Raghavan says this led Kissinger to misread the situation in December. In fact, Kissinger had radically revised his initial assessment after his October visit to Beijing. After the second visit, he reported to Nixon that China was now "afraid of giving Moscow a pretext for attack". In December, Kissinger believed that China would not intervene unless it was offered an assurance of American support. This is why he proffered an assurance of support while encouraging China to move against India in December.

Raghavan's book is intended as a 'global history'. Venturing into largely unexplored territory, he raises a number of interesting questions but at times stretches the facts to fit them into his thesis. He exaggerates the impact on East Pakistan of student movements in Europe and the US and '60s counterculture. "The counterculture of the Sixties found a fertile bed among Pakistani youngsters," he asserts.

"Rock 'n roll was an important vector for the global diffusion, including Pakistan, of the spirit of the Sixties." He presents a startling thesis: "Without the global revolts of 1968, it is exceedingly unlikely that there would have been an uprising led by Pakistani students that not only deposed Ayub Khan but also radicalised the movement for autonomy in the eastern wing." This is a far-fetched notion. The Bangladesh movement drew its inspiration from traditional Bengali culture, not an imported 'counterculture'; from the songs of Tagore and Nazrul Islam, rather than rock 'n roll. The Bangladesh student movement was inspired by the martyrs of the Language Movement, not the colourful student leaders of Britain or France.

The American edition of Gary Bass's *The Blood Telegram* carries the subtitle *Nixon, Kissinger and a Forgotten Genocide*. The subtitle of the Indian edition (*India's Secret War in East Pakistan*), evidently intended to stimulate local sales, shifts the focus away from the principal theme. Bass, a Princeton University professor of political science, focuses on US and Indian policies and the humanitarian dimensions of the 1971 war. He writes in a vivid and racy style and never fails to hold the reader's attention. The book is a thoroughly researched and damning indictment: "Nixon and Kissinger bear responsibility for a significant complicity in the slaughter of the Bengalis...this major incident has largely been whitewashed out of their legacy—and not by accident. Kissinger began telling demonstrable falsehoods about the administration's record just two weeks into the crisis, and he has not stopped distorting since."

Bass demolishes Kissinger's defence of his stand on humanitarian issues, including the refugee problem and restoration of democracy. He shows that the few initiatives taken by the US administration were not only totally inadequate but were designed to deny India a valid justification for intervention. India receives qualified approval. Bass agrees that India's military intervention was one of the most successful cases of humanitarian intervention against genocide but adds that "Indira Gandhi's government was motivated by a mix of lofty principle and brutal realpolitik". This is true, of course, but the reader may well ask whether there is even a single instance in history of a government undertaking humanitarian intervention without any concern for its national interest!

RAW AND BANGLADESH

Besides pressurizing Bangladesh through political and economic measures including water blockade RAW has been instigating and promoting separatist movements to completely destabilize and disintegrate the country. RAW has been providing training and arms to various separatist movements to create anarchy in Bangladesh. One such organisation is named Shanti Bahini. It is fighting for creating an independent state named Jhumland in Chittagong Hill Tracts region.

Historically Chittagong Hill Tracts region has been a part of ancient Bengal. Previously, it was' within Chittagong district. The Birtish government made it a separate district in 1900 A.D. Thus it was not even a district before 1900, let alone an independent state. At the time of partition in 1947, Hindu leaders including Pundit Nehru had desired that Chittagong Hill Tracts should be included in India. At that time there were 10% Hindus among the tribals of Chittagong Hill Tracts only 'Tipra: and 'Hajong' are Hindus among 11 tribes of Chittagong Hill Tracts. According to census of 1981 Tipras are 12.31% of the tribal population while the 'Hajong' are negligible. Since their demand did not have any historic, cultural or religious basis therefore it was not accepted. Indians have been eying this strategic area' even later. During sixties, Indians instigated Tipras to start trouble in the region but they did not take the bait. Thus Indian design to separate Chittagong Hill Tracts from Pakistan remained unfulfilled.

After the emergence of Bangladesh in 1971 RAW got confident that any country can be disintegrated through a series of conspiracies and covert actions. Its confidence was further bolstered after annexation of 'Sikkim' by India. RAW had hoped that due to its support during 1971 War of Independence, Bangladesh will ever remain grateful and subservient to India. RAW even hoped that in due course they will turn Bangladesh into the next Sikkim or at least Bhutan. But they were disappointed to see Bangladesh asserting for sovereignty. Indians were forced to withdraw troops form Bangladesh due to clear and unequivocal stand by Sheikh Mujib-ur-Rahman. Leter when Bangladesh joined OIC in 1974 and announced that Banglasdesh will play its due role as a memebr country of Muslim Ummah, it was not liked by Indian leaders.

After assassination of Sheikh Mujib-ur-Rahman in 1975, the new Government in Bangladesh identified itself more closely with Muslim nationalism and started following independent foreign policy. RAW was indeed taken aback by this courageous posture adopted by what they considered a satellite state. They feared that if Bangladesh grows stronger and bermes truly sovereign, it can be a big hurdle in path of their designs in the region. Therefore RAW decided to rein Bangladesh.

Indian policy makers thought if Bangladesh is allowed to prosper, the North-Eastern states of India which are adjacent to Bangladesh will be inspired further for independence. Even the insurgents of these states may get shelter and support form Bangladesh in days to come. To blockade this chance RAW thought it wise to use some ambitious Chakmas who are majority among the tnbals to destabilise and disintegrate Bangladesh. So RAW began to patronise late Manobendra Larma and others in organising a group named Janoshanghati Samity (JSS). Shanti Bahini was organized as its military wing. Training and arming of its cadres was undertaken in Tripura under supervision of Indian instructors. Thus the Indians did despite a clear clause in 'Bangladesh-India Friendship Treaty 1972' which stated that neither country would allow the use of its territory for committing any act that might cause military damage to, or constitute a threat to the security of the other country. 'Mr. Manobendra Larma met Indian intelligence agency officers after crossing the broder in 1976 (Weekly Friday: June 3, 1988). Yet when Larma realized his errors and tried to reach a peaceful settlement with Bangladesh Government, India foiled it. The Friday adds, 'The attempt to negotiate a settlement failed as the armed wing of Janoshanghati Samity (Shanti Bahini) initiated armed operations under Indian pressure in mid 1976'.

Ever-since the Shanti Bahini has been waging a bloody insurgency for creating a separate state called 'Jhumland' in Chittagong Hill Tracts region. It not only attacks security forces personnel but also targets innocent civilians who have settled in the Hill Tracts. The details of their cowardly acts are well known and we need not go into details of the same. However, the magnitude of Shanti Bahini cruelties can be assessed from the fact that it has so far (till Jul 95) killed over 25000 persons of security forces and non tribal civilians. They also have coerced some 50,000 Chakma tribals to migrate to Tripura, India where they are lodging in camps set up for them by RAW. Although majority of these so called refugees want to return to their homes, but they are not being allowed by RAW as it wants to use them in its nefarious game. These refugee camps also serve as recruiting grounds for Shanti Bahini.

RAW has created divisions and factions in Shanti Bahini to stop emergence of leadership, who may come to terms with Bangladesh. On instigation by RAW, Mr. M. N. Larma was assassinated by rivals in his party on 10 December 1983 in Kallanpur camp of Shanti Bahini in Amarpur Sub Division of Tripura state. There are reports that his assassination was arranged after RAW concluded that M.N Larma would no longer serve their interests. This allegation gets credence due to the fact that the rival leader Mr. P.K. Chakma who was involved in murder of Mr. M.N. Larma is residip.g in Agartala as a state guest under protection of RAW.

After the death of M. N. Larma, his borther Mr Santoo Larma became leader of Shanti Bahini. He started talks with Bangladesh Government to solve the dispute but RAW arranged divisioh in his party which led to failure of talks. RAW also arranged a 'World

Chakma Conference' at Calcutta, capital of West Bengal in April 1987. The four day long conference adopted several resolutions regarding separating the Chittagong Hill Tracts from Bangladesh and establishment of 'Jhumland'. The details of RAW's covert acions and involvement in Chittagong Hill Tracts insurgency have been exposed by many Chakma leaders themselves. Mr. Shamiron Dewan, Chairman of Khagrachari Local Council disclosed in a press conference that Indian Government is helping Shanti Bahini by giving shelter, MONEY, arms and training. He added that the Chakmas have no genuine eco-political cause. They are being used by RAW for achieving India's geo-political interests.

Mr. Tridip Chakma a regional commander of Shanti Bahini, for thirteen years, who later surrendered to Bangladesh authorities told the newsmen that 25 training camps are being run by RAW for Shanti Bahin guerillas' in the Indian states of Tripura and Mizoram. Some separatists also received training in Deradoon, Uttar Pradesh. He said that Shanti Bahini compels tribal people to migrate to India and stay in refugee camps. Nine refugee camps have been set up at Sabrum, Shikichari, Boalpara, Kadamtolly, Daek, Barachara, Ralma, Trimakha and Ratnanager. He alleged that India continuously keeps propagating against Bangladesh regarding Chittagong Hill Tracts in a bid to make it an international issue (Bangladesh Observer: September 2, 1994). A Nepalese fortnightly 'The Spot Light' in its April 29, 1988 issue alleged that a group of Indian policy makers inimically uses Shanti Bahini to launch destructive activities inside Bangladesh. It said that the Chakma insurgents flee to their bases in India after their operations. It added that on several occasions the Chakma insurgents left behind huge India made arms and ammunition including sub-machine guns, carbines, SLRS, sten guns, 303 rifles while fleeing to India on being chased by Bangladeshi security forces.

In an interview with the 'Asia Age' Mr. Farooq Subhan, the' Ex-High Commissonner of Bangladesh to India, who is otherwise considered to be soft towards India alleged. 'The Government of Bangladesh felt that the main support for the Shanti Bahini was being provided by the Research and Analysis Wing in India (Daily Star: Decemebr 20, 1994). Indian propaganda experts spread news of human rights abuses in Chittagong Hill Tracts. RAW sponsored video films which falsely show brutalities by Banagladeshi Security Forces in Chittaogng Hill Tracts have been sent to the donor countries to suggest stoppage of aid to Bangladesh for violating human rights.

Question arises why India is helping this secessionist movement in Bangladesh. The answer lies in the following:

a. India does not want a stable, prosperous and strong Bangladesh. An ongoing insurgency will be a constant drain on Bangladesh's economy. Indians hope that by such tail twisting they can render Bangladesh to the status of a client state.

b. India wants to keep Chittagong Hill Tracts issue alive to pressurize Bangladesh and prevent her from adopting independent foreign policy. India wants that Bangladesh must align itself with Indian camp. The Chakma. Refugees and continuing insurgency are the tools in their hands to punish Bangladesh, should it ever falter in this regard.

c. Creation of a separate state in Chittagong Hill Tracts will weaken Bangladesh who can then be forced to abandon its claim of sovereignty and reduced to the status of Bhutan or even turned into Sikkim.

d. RAW maliciously propagates that insurgent groups of seven North-Eastern states use Chittagong Hill Tracts region for shelter and passage of arms and ammunition. They propound that setting up an Indian dominated state in Chittagong Hill Tracts will result in drying up of supply sources to militants and thus will cause serious setback to revolutionary movements in these Indian states (Assam, Tripura, Manipur, Mizoram, Nagaland, Arunachal Pardesh and Meghalaya). In other words according to RAW's theory, for consolidation of Indian North Eastern region, separation of Chittagong Hill Tracts from Bangladesh is essential.

e. Indian dream of getting transit through Chittagong Port to North Eastern states can be realized with establishment of a pro-Indian state in Chittagong Hill Tracts.

RAW is actively pursuing its agenda to disintegrate Bangladesh. They always remain in search of an issue for fulfilling their objective. Since there is no division amongst masses with regards to language, ethnicity and culture, RAW has been relying upon religious differences to initiate and develop separatist movements in the country. Bangabhumi is the proposed name of a homeland for the Hindus who had migrated to India from East Bengal in the past. It is stated that one and half crore Hindus uprooted from the basins of Rivers Padma, Meghna and Madhumati who were forced to leave the area on account of economic deprivation and communal riots, now claim to have a right to settle in their own motherland. The state is planned to comprise of the districts of greater Khulna, Jessore, Kushtia, Faridpur, Barisal and Patuakhali of Bangladesh.

The start of Bangbhumi movement was announced on August 15, 1977. Its office is located in Ramlal Bazar of South Calcutta. The organisation styled itself as 'Nikhil Banga Nagarik Sangha' (All Bengal Citizen's Organisation). Mr. Sri Partha Samanta is the President and Mr. Kalidas Vaidya is the General Secretary of the Organisation. On March 25, 1982 'Nikhil Banga Nagarik Sangha' officially declared the formation of 'Bangabhumi' inside the territory of Bangladesh. Besides forming the 'Nikhil Banga Nagarik Sangha', RAW has also set up some paper organisations i.e., 'Bangla Liberation Organisation' styled on the pattern of 'Palestine Liberation Organisation' and 'Liberation Tigers of Bengal'on the pattern of Liberation Tigers of Tamil Elarn.

Before we expose RAW's support to the Bangabhumi activists, it is pertinent to analyze rationale of their demand. It is a known fact that after partition in 1947 Hindus had the option to settle either in Pakistan or migrate to India. While many Hindus stayed back in the then East Pakistan, many others migrated to India. Likewise many Muslims left India and migrated to Pakistan. Such migrations were final and cannot be revoked on any pretext' now. The question arises that on what pretext RAW wants to create a separate homeland for Hindus who had migrated to India voluntarily at their own will from East Pakistan (now Bangladesh) in 1947. During the united Pakistan, and even prior to 1977, such a preposterous claim was never raised. Such an attempt amounts to interference in the independence, sovereignty and regional integrity of Bangladesh.

The total area of proposed Bangbhumi comes to 20 thousand square miles which is more than one third of the total area 'of Bangladesh. The border of Bangabhumi runs in the north along River Padma, in the east along River Meghna,in the west along India-Banglade'sh border and in the south along the Bay of Bengal. The declared capital of the s.tate is at Samantanager (Mukti Bhaban or Freedom House) which has no apparent existence. According to reliable sources Samantanager is nom de plum for Bangabhumi HQ established in a house named 'Sunny Villa' in Bhabanipur, Calcutta. This house is owned by Mr. Chittaranjan Sutar, a famous RAW agent. 'Sunny Villa' had once played a pioneering role in organising activities for disintegration of Pakistan and creation of Bangladesh. That very Sunny Villa is again active and is being utilised this time for disintegr,ating Bangladesh. The reason is that Bangladesh has declined to bow before Indian designs thus it ought to be taught a lesson.

The flag of proposed Bangabhumi state comprises of one third green colour and the two third of saffron colour. In the middle of the saffron colour is the white sun. On the top of the stand is' a circle. The National Anthem is the falrlous song of poet Dijendra Lal Roy: 'Dhana Dhanye Pushpe Bhara, Arnader Ei Basundhara'. (This land of ours filled with crops and flowers). The National Symbol is depicted by a green circle in another saffron coloured circle with word 'Sri', (Symbol of beauty and prosperity) written in the saffron circle. The state policy proposed to be adopteds non-communal and humanism. The name of propsed radio station is 'Bangabani' keeping in line with 'Akashvani' and its' call s-ign is 'Joy Banga' in line with 'Jai Hind' of India.

An armed wing of the organisation named 'Bangasena' has also been formed. The Commander of this armed force is Mr Kalidas Vaidya. The main organiser and trainer of Bangasena is Arun Ghosh. The only condition for enlistment in 'Bangasena' is that the recruit should be a resident or had been a resident or born in East Bengal. If either of the answers is in affirmative then he is considered eligible to be enlisted in 'Bangasena'. The daily

Anandabazar Patrika of Calcutta, while writing about ''Bangasena' reported, 'The majority mempers of this Bangasena which is a voluntary force lack any plitical consciousness. They are utterly communal in their outlook, semiliterate and very emotional (Daily Anandabazar : September 13, 1989).

RAW has deliberately recruited Hindu fanatics and extremists in the organisation so that it is easy to get any mischief done from them. However, the experts say that though Bangasena apparently comprises of riff raffs of Calcutta, but many former officers and personnel of the Indian Armed Forces are also working behind the scene.

The supporters and activists of Bangabhumi had threatened to enter into Bangladesh from India many times in the past. But all their attempts failed. In Bangladesh they had circulated posters and publicity material printed in India through agents of RAW. Many such posters, leaflets and pamphlets were seized by police from Naryanganj, Kushtia, Jessore, Khulna and Barisal.

Besides, police also seized many so-called flags of Bangabhumi which the agents of RAW had smuggled from India for hoisting on Government buildings. However, the RAW agents did manage to hoist Bangabhumi flag at three places in Tala Thana of Jessore. But the local people and the police immediately seized the flags. Bangabhumi leaders also sent letters to the Deputy Commissioners and other senior officers of the area threatening them to ex-communicate with Government of Bangladesh by March 23, 1985 and owe their allegiance to the so called Bangabhumi Government failing which they will be killed alongwith their families.

To create harassment amongst Bangladeshis of this region and to attract international news media, RAW provoked its agents to launch an,attack on the office of the Bangladesh Deputy High Commission in Calcutta (April 11, 1994). The agents of RAW justified their action by saying that since Bangabhumi has come into being, therefore, they were justified to own and occupy Bangladeshi Mission in Calcutta. They attacked building of Bangladesh Deputy High Commission in presence of Indian police, entered into the premises of the Mission and hoisted the flag of so called Bangabhumi.. RAW had enacted similar drama through its agents at the office of the Bangladesh High Commission in New Delhi on August 15, 1984.

RAW wants to create disturbances, terrorism and communal trouble in Banladesh on the plea of Bangabhumi movement. The agents of 'Bangasena' who are armed with modern weapons are looking for an opportunity to infiltrate into Bangladesh. Thereafter they will create such conditions in the region through blood letting and rioting that India may be justified to march in its regular Armed Forces into the area to establish 'peace and order'. That is how Indians are ultimately planning to separate western part of Bangladesh. However, this is a long term plan. For the time being, the Bangasena is being nurtured by RAW to function as a pressure group to compel Bangladesh to adopt a Pro-India policy. Nevertheless they are being prepared for the fore mentioned greater objective to be pursued in due course. After all RAW is not an orphan house that it will keep spending crores of rupees on such organisations without any ultimate aim in view.

There is no end to RAW's anti-Bangladesh activities. RAW is relentlessly pursuing the agenda to harass and subjugate Bangladesh. A net of conspiracies is being regularly dished out from its research cell for this purpose. As part of its strategy to destabilize Bangledesh, RAW is promoting a mushroom growth of organisations involved in anti-Bangladesh activities covering all fields including political, economic, cultural and terrorist activities. Some of the RAW, partronized organisations in India are Bangabhumi Movement, Bangladesh Udvastu Kalyan Parishad (Bangladesh Refugee Welfare Organisation) and Bangladesh Mohajir Sangha (Bangladesh Mohajir Organisation) etc. The first organisation i.e., the 'Bangabhumi Movement' is demanding establishment of a separate homeland for the Hindus in Bangladesh comprising the district of greater Faridpur, Kushtia, Jessore, Khulna, Barisal and Patuakhali. It is threatening to launch an armed movement for realizing its objective.

The second organisation is composed of Hindus who at the time of partition in 1947 had migrated to India on their own after disposing off their movable and immovable properties. Now they are complaining that they were forced to leave the then East

Pakistan and they claim to return to Bangladesh. They also want to get back their properties.

The third organisation, 'Bangladesh Mohajir Sangha' claims to represent Muslims of Bangladesh who had entered India illegally in the past under various compulsions and now they want to come back to their motherland i.e., Bangaldesh. The organisation is a front for the Indians to help in push back of so called illegal migrants from Bangaldesh. In this chapter we shall briefly explain the origin, activities and objectives of Mohajir Sangha.

Bangladesh Mohajir Sangha was formally launched by its so called leaders on February 12, 1991 by holding a press conference at Calcutta press club. The dramatic emergence of this organisation was given wide coverage by radio, television and newspapers in India. Since radio and television are government controlled, so it can be easily and safely presumed who the actual sponsors were. The statement of Mohajir Sangha broadcast by All India Radio and telecasted by Dooradarshan says, "We are the Muslims of Bangladesh who had illegally entered India in thousands. Now being fed up of living in India we appeal to World Human Rights Commission for our repatriation"(Monthly Kalam : February 1992).

The question arises that how was it possible for such a large number of Muslims to illegally cross the borders without the Indian border security forces even noticing them. More so how did the Hindus allow them to settle in India for such a long time and that they were never arrested or jailed. Now all of a sudden they emerge from no where and start demanding repatriation to Bangladesh. The president of Mohajir Sangha is a rickshaw driver named Raisuddin Ahmed. He is assisted in his anti-Bangladesh activities by Chittaranjan Sutar. Intrigued by the activities of Mohajir Sangha and the unprecedented interest of Indian Government and media, a correspondent of Indian Monthly 'Kalam' conducted detailed inquiry to ascertain how this organisation was established and FINANCED. 'Kalam,' in its issue of February 1992, published a lengthy report comprising of 14 pages regarding Mohajir Sangha and Udbastu Kalyan Parishad. The correspondent of 'Kalam' also held two intimate interviews with the President of Mohajir Sangha Raisuddin Ahmed.

As a result of the investigation the magazine correspondent was able to trace the true details and addresses of this organisation. The President Mr. Raisuddin lives in a hut in the slum area on the banks of Ganges-Kashipur in Calcutta. He used to drive a hired rickshaw of one Nimai Gosh. He claimed that he originally belonged to a village in the District of Jessore in Bangladesh and had illegally migrated to India some years ago. However, he could not substantiate his claim about being Bangladeshi national. He is illiterate. The magazine noted that it is virtually impossible for an illiterate person like Mr. Raisuddin to draft such sophisticated leaflets containing appeals in the name of human rights supported by English quotations, leave alone getting them printed and distributing amongst international community. Raisuddin admitted during interview that Hindu babus make these arrangements and also provide FINANCE for the organisation.

Actually it is RAW that FINANCES its members. Raisuddin now does not drive the rickshaw and mostly remains busy in meetings and processions. According to the publicity leaflets the head office of Mohajir Sangha is shown at 5/1, Khagendranath Chattejee Road, Calcutta-2, but actually the headquarters is located at Bhabanipur in a house named 'Sunny villa'. The long time resident of palatial Sunny villa is RAW's most trusted, faithful and successful agent Chittaranjan Sutar.

It is common knowledge that in the year 1970 the Indian intelligence had handed over this multi-storied building to Chittaranjan Sutar for organising operations against erstwhile East Pakistan. From this villa were initiated all strategies, plans, instructions, coordination and supervision, for disintegrating Pa~istan. Chittaranjan Sutar originally belonged to Barisal, Bangladesh. He and his two other colleagues Kalidas Vaidya and Nirod Majumdar were engaged since 1952 in activities aimed at disintegration of Pakistan.

He migrated to India in 1969.After emergence of Bangladesh as an independent state Chittaranjan Sutar returned to Bangladesh with pomp and show. During 1973 elections he obtained nomination from Awami League and contested elections from Swarupkathi constituency. He won the election and became member of Bangladesh Parliament. After assassination of Sheikh Mujibur Rahman in 1975 he went back to Sunny Villa in Calcutta and stationed himself there for good.

Chittaranjan Sutar is not satisfied with the disintegration of Pakistan and consequent emergence of Bangladesh. His aim, as, directed by RAW, is to gradually integrate Bangladesh with India. But due to the changed scenario he could not implement his vicious designs. RAW drew a new strategy; that of partitioning Bangladesh. Chittaranjan Sutar is involved in implementation of the new game plan. As a part of this scheme the plans were drawn to constantly keep on harassing Bangladesh on different issues. Mohajir Sangha was one such weapon to harass Bangladesh. Through Mohajir Songha, Indians wanted to convince the rest of the world that illegal of Bangladeshis into India was not merely propaganda but it was really happening. Thus creating. Justification for eviction of lakhs of Muslims from Indh{where they had lived for generations. Such mass exodus into an already over populous country like Bangladesh will spell doom for the country. Besides, under the garb of these so-called Mohajirs, RAW would infiltrate into Bangladesh its trained terrorists. India can provoke the so-called Mohajirs to claim a separate homeland and may even instigate them to wage subversive actions against Bangladesh culminating into a limited war.

Meanwhile in different Indian states lakhs of Muslims have been discriminated by not enlisting them as voters. They have been denied this right despite the fact that they are Indian nationals and are residing in India since generations. Some who were earlier enlisted as voters and had been casting their votes "in the past elections have also been denied entry in new election roles. The main purpose of this scheme of RAW is to establish that lakhs of Muslims had illegally infiltrated into India from Bangladesh and prepare a strong base for th~ir repatriation.

During the visit of Bangladesh Foreign Minister to India in 1991 RAW arranged a mass sit-in and hunger strike by the Mohajir Sangha. A leaflet circulated at the occasion by Mohajir Sangha read, 'foreign aid misappropriated, sovereignty at stake, basic principles of constitution framed after independence abandoned, a few people have usurped all the wealth of Bangladesh and have destroyed the economic base of B"angladesh, the labour class has been rendered into beggars.Their demands included holding trial of killers of Bangabandhu, rehabilitation of labour and restoration of the human rights. A memorandum was also submitted to the Bangladesh Foreign Minister on this occasion.

From the language of the above mentioned leaflet the objectives of the Mohajir Sangha can be easily guessed. Needless to mention that the memorandum presented to the Foreign Minister was given wide publicity through the government media i.e.; All India Radio and Doordarshan. Copies of the memorandum were also sent to the foreign embassies and missions in New Delhi. RAW arranged wide media publicity for the activities of Mohajir Sangha. Incredible editorials and articles were published in newspapers like Ananda Bazar Patrika, Bartaman, Jugantar, Statesman, Sunday, Swatikatha, Ganashakti etc.

All these papers gave wide coverage to the statements of Mohajir Sangha. The papers expressed their indignation at the failure of the Government in identifying the infiltrators and arranging their eviction. Curiously enough Mr. Raisuddin, the man who confessed himself of illegally infiltrating into India was not arrested. The media did not utter even a single wort! to say under what right of the law such an illegal infiltrator had started an organisation in India, held a press conference, took out a procession with the slogan of 'Allahu Akbar', held mass hunger strike and how their activities were given wide coverage through government media i.e. Doordarshan and All India Radio.

The papers did not raise any question about it, because they knew that Bangladesh Mohajir Sangha was, in fact, their weapon for Bangladesh bashing. They understood that this weapon should not be blunted, rather it has to be kept sharpened ever and always. The establishment of organisations such as Mohajir Sangha is proof enough of the fact that RAW and in turn the Government of India is actively and continuously harassing Bangladesh. The most unfortunate part, however, is that no effort is made in Bangladesh to expose these Indian designs. Our common people and even the educated class in the urban areas do not know any thing about it. Experienced observers are of the view that RAW has strong influence over important state organs and individuals who connive with RAW. Unless RAW's agents in various echelons of Bangladesh are eliminated the country's interests and sovereignty will remain at the risk of being compromised.

'Push Back' is the common term used to describe Indian scheme to send hundreds of thousands of Indian citizens into Bangladesh. India claims that these' people had illegally entered into -India from Bangladesh and thus should be 'pushed back' to their country. At occasions Indians did try to push back some people, but prompt action by Bangladesh authorities thwarted their efforts. However, a critical analysis indicates that the issue has been coined by RAW with long term ulterior motives. These include :

a. To raise bogy of an imaginary threat of large scale infiltration from Bangladesh into India and thus arousing public sentiments against Bangladesh. RAW believes that to have real or imaginary enemies is necessary for maintaining national integration.
b. A large number of agents can be penetrated and permanently settled in Bangladesh by RAW using the cover of push back. These agents can then work for disintegration of Bangladesh from within.
c. The issue can be used as an excuse to wage war against Bangladesh in the name of securing rights of large number of Bangladeshi refugees. Ever since the creation of Bangladesh various Indian Governments have been endeavouring to whip up the issue. Dr. SS Bindra in his book 'India and Her Neighbours' (page 166) writes, "During Sheikh Mujib's regime, the Government of India protested to Bangladesh that large number of Bangladeshi nationals had entered into West Bengal.

The Bangladesh Government denied that any of their nationals had crossed over into India. They argued that these IMMIGRANTS were actually 'Indian nationals' who had travelled to Bangladesh when the freedom struggle in Bangladesh was going on. But after some time they returned to their own country. Commenting on the issue The Daily Statesman wrote on 13 October 1974, "It is possible that a sizeable number of 1947 refugees had returned to their original homes in 1971.

But Bangladesh Government considers them aliens. Many of them returned to India who are now considered as Bangladeshi nationals by India. It is obvious that Bangladesh cannot accept such an absurd claim. The migration in the wake of partition of 1947 was final. No body has the right to reclaim his original nationality after 48 or 24 years. Actually after 1971 Hindus of West Bengal had wanted to regain their lands and property in Bangladesh which they sold while migrating to India in 1947. They 'thought Bangladeshi Muslims would not dare to face these Indian Hindus because of Indian pressure. With these hopes they had returned to Bangladesh after the surrender of Pakistan Army. Bangladesh authorities could not check this exodus because no organized authority existed in the country at that time due to the war. However, these people were never officially recognized as Bangladeshi nationals. When these people returned to India, in fact, they returned to their own country. Thus Indian claim that these people are Bangladeshis, is totally untenable.

The Indian propaganda about infiltration of Bangladeshis into India has continued unabated. It is used as a means to divert attention of the people from internal problems of the region (West Bengal and North Eastern states). Recently Indian Home Minister Mr. SB Chaven stated in an interview to BBC that hundreds 'of thousands of Bangladeshis are entering India illegally everyday (Daily Janata : May 14, 1995).

One should ask India that how it was possible for such large numbers to cross over the border in the presence of elaborate network of Indian border, security agencies? There are 882 Indian border outposts along the Indo-Bangladesh border. In comparison Bangladesh has only 462 border outposts. '

Mr. Tapan Sikdar, President of West Bengal BJP unit told in an interview to Monthly Kalam, January 1994, 'There are 7.5 million Bangladeshi Muslim in West Bengal and in total 20 million throughout India. Again one is at a loss to understand how such a large scale, migration could take place without being noticed and checked at the borders.

Actually RAW wants to inflate figures to pressurize Bangladesh and in the process obtain her commitment to take back at least token numbers. In this regard RAW had also organized 'Muhajir Sangha', a movement of so called Bangladeshi in India who claim to go back to Bangladesh.

An interesting point to note is that while Central Government is crying hoarse about illegal, the state Governments of West Bengal and Assam have been stating that no illegal immigrants are staying in their states. Similarly before assumption of power by BJP Government in Maharashtra, there was no talk of illegal Bangladeshi Muslim immigrants in Bombay. The dichotomy in stands of various state Governments clearly indicates the falsehood of the claim. While BJP is fuelling the issue due to its communal out look, the Central Government has other ulterior motives. Foremost is the diversion of attention of the people from their grievances and giving them a plausible justification for their sufferings.

Let us first take the case of West Bengal. People of West Bengal feel that they have been exploited and deprived of their due share. In Calcutta 68% houses and 80% industrial units are owned by non-Bengalis. Out of 80 jute mills in West Bengal, 79 are owned by 10 non-Bengalee families. 60% of the labour force in Calcutta are non-Bengalis. Almost all top positions in industrial sector are held by non-Bengalees.

The Bengali language has virtually become alien in its own home. Particularly in urban areas Hindi has become the lingua franca. Hindi films are screened in most of the cinemas. Sign boards of Government offices as well as number plates of motor vehicles are forbidden to be written in Bangla language. Calcutta which was the most developed city of India at the time of partition has turned into virtual slum due to constant neglect and lack of development work. Similarly lack of industrial development added to people's economic miseries. As a result many people have migrated to other states and cities in search of jobs. These include Muslims as well as Hindus of West Bengal. While the Hindu migrants from West Bengal are accepted by other states, the Bengali Muslims of West Bengal are discriminated against and labeled as

Bangladeshis. This is indeed display of sheer prejudice against Muslims in so called secular India.

The exodus of Muslims from West Bengal is evident from the fact that while at the time of partition, population of Muslims in West Bengal was 45%, now it has shrunk to 29%. Where have the others gone? Certainly there have been no reports about such large scale migration from India to Bangladesh. Situation in Assam is no different which has been systematically deprived of its natural wealth by Central Government. As stated by Mr. AK Das in his book Assam's Agony', 'There has been serious lack of interest on part of the Government for developing communication, power, education, health, social welfare, agriculature and industries in Assam'.. Mr. Das further writes-in his book,' The Government of India neglected the North East Region in the development of industnes: in the public sector. The Government has made about 20 percent of its in Maharashtra and Gujarat as compared to only 5 percent in this region

The North East Region has only 21 large and medium scale industrial units out. of 15,000 in India. India has 1,750,000 small-scale industrial units with of Rs 12,000. million. But this region has only 3,000 units with a paltry of Rs. 150 million. The industrial employment is only 4 percent as compared to 20 per cent in India'(Assam's Agony, Page -233).Assam produces 58% tea of total Indian production and 30% of total tea production of the world. But only one - ninth of the total tea is auctioned in Gauhati. Rest is auctioned in Calcutta and other places. Agriculture goods'worth 30,000 million rupees are taken llway from Assam to other parts of India annually. 60% crude petroleum of total Indian production is extracted in Assam.

Similarly large quantity of natural gas is also extracted in Assam. But Assam's share in Government is not proportionate to its contribution. 60% Government employees in Assam are outsiders. The share of Assamese working in Assam in railways is 14%, telegraph and postal services is 10% and, plywood industries lis 15%. The remaining employees in these departments are all from other states. In such a situation, the growth of discontentment is natural. Therefore separatist movements have grown in Assam and other North Eastern states. The seriousness of the situation was acknowledged by no body less than Indian Army Chief himself who recently stated that insurgency situation in North East region was next only'to Kashmir.

However, instead of taking appropriate corrective measures' to obviate people's sense of deprivation, the Central Government created the bogy of infiltration from Bangladesh as an excuse for poor development of the region. Initially some people in Assam bought the idea.

'Assam Gano Parishad' adopted it as election slogan in 1985 and came to power. However, gradually truth has dawned, on the people. They now understand Central Government's machinations. That is why Mr. Saikia, the present Chief Minister of Assam categorically stated recently that there were no illegal Bangladeshis in Assam.

On the other hand Central Government ministers still keep harping on the same old theme. Let us also discuss, the illegal issue propounded by Maharashtra Government of BJP. Reportedly many Bengali speaking Muslim families have been uprooted from Bombay, but independent observers noted that all of them belonged to West Bengal. AFP after sampling 100 such families reported that these people had come from West Bangal and not from Bangladesh.

However, Maharashtra Government is bent upon calling them Bangladeshis as they happen to be Bengali speaking Muslims. Do they want.to say that only Hindus have right to belong to West Bengal? It is quite apparent 'from the above discussion that so called 'Push Back' is a ploy used by India to pressurize Bangladesh, besides diverting attention of the people of aggrieved states. RAW cleverly and wickedly uses the issue to raise anti-Bangladesh hysteria whenever necessary for internal or external reasons.

RAW' believes that by such tactics it can keep Bangladesh under pressure and hence push it to total subjugation. Moreover, use of the issue by the Indians in combination with their bigger game plan i.e. invasion of Bangladesh claiming that their national interests have been endangered due to large scale alien, cannot be ruled out.

Hindu Organizations

RAW has sponsored the establishment of a large number of Hindu organizations in Bangladesh for a variety of purposes to include; rendering support to pro-Indian political parties, to give shelter to Indian terrorists, to divide people on various issues, to spread communal and class hatred and finally to create favourable conditions for merger of Bangladesh into India. Common people are not fully aware of all such organizations and their activities. They mostly know only about some political groups and cultural forums who are widely reputed to be pro-India. However, there are many other organizations that are working against national interests and it is important that their activities should be brought to the notice of our people. Of particular interest in this regard are the RAW sponsored organizations of minority community.

The Hindu communalists have not till today reconciled with the partition of India in 1947. They may be living in Bangladesh but their loyalty and allegiance are with India. Many of their communal organizations are working to undo the partition. In other words they are working against the integrity of Bangladesh. RAW is their main financier. However, these organizations also get and other support from various extremist Hindu parties of India including BJP, VHP, Shiv Sena, Bajrang Dal etc. The only overt organisation of minorities is Hindu Buddha-Christian Oikya Parishad. Other organisations work so discreetly that not much is known about their activities out side their own community.

Names of some of these organisations are :

a. Bangladesh Jatiya Hindu Samaj Sanskar Samity (BJHSSS).
b. Pranab Moth Shevasharam (It was re-organised from defunct Bharat Shevasharam).
c. Bangladesh Sant Mohamandal.
d. Bangladesh Hindu Foundation.
e. Vishwa Dharma Shanti Shammelan Bangladesh.
f. Bangladesh Chaitanna Shanskritic Shangh.
g. International Society for Krishna Conciousness(ISKcoN). Its head Office is in Mayapur, District Nadia, West Bengal, India.

h. Ananda Marg(Daily Meillat : June 7, 1992).

Besides above, a number of other small organisations also exist which virtually work as the Bangladesh branches of communal extremist Hindu organisations of India. These organisations outwardly profess to be doing humanitarian work i.e. alleviation of illiteracy etc. but actually these are dedicated for doing communal work.Some minority organisations also publish newspapers and magazines etc. A Hindu magazine 'Samaj Darpan' is published by Bangladesh Jatiya Hindu Samaj Sanskar Samity (BJHSSS). Its office is located at Vholagiri Ashram, K. S. Lane, Dhaka. The cover page of their magazine.is printed in Calcutta, India, due to the lack of photographs of Hindu gods and goddesses in Bangladesh.

It is believed that the entire expenses on printing and publication of this magazine are borne by RAW. Reportedly BJHSSS also published booklets - and leaflets supporting Bangabhumi movement and Akhand Bharat. The head of BJHSSS, Mr. Shiv Shankar Chakarvarti, who is also the editor of Samaj Darpan is reported to have close links with Indian Vishwa Hindu Parishad (VHP). He is also a supporter of Bangabhumi movement. His close associates include Mr. Bijoy Nandi, head of India Shevasharam, Mr. Vinod Bihari Chowdhury, President Bangladesh Hindu FOl1ndation and Mr. RN Das Gupta, Secretary General Vishwa Shanti Shammelan.

According to the Daily Meillat, June 10 1992, Mr. Bijoy Nandi told in a meeting of Bangabhumi activists at Bangaon, West Bengal that in 1972 the then Government of Bangladesh had given permission to the organisation (India Shevasharam) to work in Bangladesh due to personal influence of Mrs. Indira Ghandi. After the assassination of Sheikh Mujib on 15 August, 1975 all officials of the organisation fled to India. Later they returned to Bangladesh and formed another organisation named Pronab Moth Shevasharam.

The Daily Meillat, 8 June, 1992 wrote: 'Mr. Shiv Shanker Chkarvarti holds positions in different organisations. He is a member of executive committee of VHP in India and the chief co-ordinator of VHP in Bangladesh. He is also member of executive committee of Nepal based Hindu Mahasangha, President of Hindu

Samaj Sanskar Samity, Editor of the Samaj Darpan and above all, chief cashier of Apex Group of Industries. He earned much recognition from RAW for working as a sincere local agent. His magazine 'Samaj Darpan' is notorious for making massive anti-Bangladesh and anti-Muslim rhetoric. He spreads message of communal hatred and continuously propagates that "Hindus are being persecuted in Bangladesh. At times his writings are so provocative that these amount to inciting Hindus to take up arms for getting their so called due rights. Mr. Bijoy Nandi, who is the President of pronab Moth. Shevasharam, in his speeches, launches ferocious attacks against Bangladeshi Muslim community. He has been inciting and urging Hindu youth to join Bangabhumi movement.

The Daily Meillat, June, 10, 1992 quoted him saying: 'You (Hindus) laid down your lives to secure independence of India. Your forefathers fought against the aggressor Pathans and Mughals to save the temples ofIndia. Your forefathers also fought to safeguard our religion and civilization from the atrocities of Aurangzeb. My dear fellow beings! the youth of Bengal and India! Come and join Bengasena and attack Bangladesh to establish Bangabhumi in six south western districts (Faridpur, Kushtia, Jessor, Khulna, Barisal and Patuakhali). Let the UNO and the USA be damned. Let Bangladesh know our strength. Let Indian Muslims understand that Hindus are still alive and powerful! The biggest Hindu organisation in Bangladesh is Bangladesh Sant Mohamandal'. Its members are spread all over Bangladesh. It has 267 -member Central Committee. It also has a 10-member Presidium Council.

Bangladesh Sant Mohamandal has close links with the Communist Party of Bangladesh and 5 Parties Alliance who are blamed to be pro-India. It tries to control rural politics of Bangladesh through Hindu voters. It also co-ordinates electoral policy of Hindu voters in collaboration with a large national secular political party. The organisation also serves as a recruiting ground for RAW.

All— these organisations are engaged in relentless propaganda against Bangladesh. They spread misinformation, rumours and false reports to paint Bangladesh in poor light. On the contrary

India is depicted as a mighty super power. The aim of such propaganda is to convince people that existence of Bangladesh as an independent country is not viable. They quote fictitious figures to prove their point and argue that merger of Bangladesh in Indian union will be in the larger interest of the Bangladeshi people. Members of these organisations are willing and dutiful supporters of RAW in its campaign against Bangladesh.

They are very motivated for the cause of 'Akhand Bharat' and in many cases work without any FINANCIAL or other rewards. They are hardened Hindus who are willing to render any sacrifice. They provide all types of support required by RAW in its operations. From managing safe houses to the conduct of terrorist activities, these volunteers are willing to accomplish any task. Due to their discipline and dedication it is difficult to penetrate these organisation. Hence they are able to do their activities in total secrecy. RAW is indeed lucky to have a large community of hard core and motivated Hindus in Bangladesh who is ready to do anything for the cause of Akhand Bharat. This situation can indeed be compared to the support enjoyed by MOSSAD from all Jewish people.

These extremist Hindu organizations are posing grave threat to country's independence and communal peace. Due to their extreme secrecy and discreet working, very little is known about their activities. Taking advantage of the situation, RAW is sponsoring more such religious organizations day by day. It is time that our Government realizes the threat from such communal organizations and necessary steps are taken to regulate the working of these so called religious organizations.

ARMED FORCES

Nobody now refutes the proposition that India in engineering dismemberment of Pakistan in 1971, had really aimed at taking a step forward to re-establish united India. In view of her said goal, India ensured right from 1971 that Bangladesh does not become a strong country. India wants that Bangladesh should remain a nominally independent country like Bhutan and Sikkini which could be occupied later by India when she desires. In furtherance of the said aim India had forced the 1971 Interim Government of Bangladesh to sign a treaty of subservience. One of the clauses of the treaty was that Bangladesh will not raise any Armed Forces of its own. Indians had told the then Interim Government in exile that Bangladesh does not need an 'Army' because she is surrounded on three sides by India and the other side by natural sea. Since India was a friendly country, therefore, there will be ne danger of an aggression by her. If any other power resorts to an aggression against Bangladesh, Indian Army would defend Bangladesh, they advised that for internal security Bangladesh should raise a militia force.

That is why Rakhi Bahini was formed after creation of Bangladesh. When Sheikh Mujib still decided to raise national Army it was opposed by India. However, he decided to go ahead with the raising of the Army despite India's opposition. Thus a small Army was raised comprising of soldiers of Pakistan Army who had taken part in freedom struggle. Later most of the army personnel repatriated from Pakistan in 1973 were also enrolled in Bangladesh Army.

RAW apprehended that if Bangladesh succeeds in building powerful Armed Forces, it will become difficult for India to realize her dream about merger of the new country into India. The Indians knew well about the courage and skill of the Bengali army men. The experience gained during -the liberation war was an additional asset for Bangladesh Army. Indian policy planners concluded that even small Bangladesh Army might in future pose a threat to their aggressive designs.

Indian designs against Bangladesh Army were evident from the very beginning. They did not leave any military equipment captured from Pakistan Army in the then East Pakistan. In this regard the Daily Meillat reported, "The Indians took away the abandoned military hardwares (tanks, artillery, guns etc.) worth Taka 80,000 million of the Pakistan Army's Eastern Command in order for materialization of their goal i.e. not to permit Bangladesh Army to grow strong(Daily Meillat : May 26, 1993).When Indians realized that Sheikh Mujib is determined to raise an Army in defiance of the treaty signed by Interim Bangladesh Government and India, they encouraged a dissident party to take up the issue of raising a people's Army or 'Gano Bahini'. The dissident leaders started propagating that Bangladesh does not require a regular professional Army. They labeled the Army as God's Khasi (castrated of God). They organised so-called 'Revolutionary Soldier's Council' to realise the goal of 'Scientific Socialism'. It was actually formed to create division and disunity among the rank and file of the Armed Services.

Soon after the change over of 1975 some cadres of Gono Bahini attempted to create chaos and lawlessness in the country. RAW also stage-managed fake kidnapping in Dhaka of Mr Samor Sen, the then Indian High 90mmissioner in Bangladesh. The aim was to create plausible excuse for India to move in Indian Army into Bangladesh. However, the 'kidnapping' drama did not go too far and thus Bangladesh was saved from direct Indian aggression. After coming to power the late President Ziaur Rahman re-organized the Armed Forces. He gave them an honourable position in the society. But RAW did its best to obstruct him. That is the reason for so many coups staged against Ziaur Rahman. He was killed in the 19th coup attempt. Although apparently the coups were aimed at ousting Zia but the ultimate goal was to weaken or liquidate the Armed Forces.

RAW engineered conspiracies to divide rank and file of the Bangladesh Army in various groups. The division between 'Freedom Fighters' and 'Repatriated' was fuelled to pave way for internal strife and disintegration. RAW has also been endeavoring to project Bangladesh Army as'anti-people'. During the peak period of anti-Ershad movement, RAW spread a rumour that the Army was supporting Ershad. The rumor was aimed at damaging the popularity of the Armed Forces of Bangladesh.In order to protect country's independence, sovereignty and territorial integrity there is no substitute to adequately strong Armed Forces. If Sikkim had an effective and strong Army she would not have been occupied and annexed by India. On the other hand Srilanka could withstand Indian pressure because she has her own effective Army.The Maldives, on the contrary, had to seek Indian Army's help to drive out Indian henchmen from Maldives, because she had no Armed

Forces of her own. Bangladesh must have its own Army not only to thwart aggression from abroad but also to quell internal troubles and rebellions. The credit goes to valiant Bangladesh Army for checking nefarious designs of Shanti Bahini. If Bangladesh did not have an effective Army, Indian supported Chakma insurgents might well have succeeded in dividing the country. Similarly Indian Army would have marched into Bangladesh on any flimsy pretext. The need for strong Army is further highlighted by growing internal threats supported by- India in form of secessionist movements in Bangladesh in the name and style of Jhumland, Hinduland (Bangabhumi) and Garoland etc.

However, the RAW sponsored henchmen have been unleashing constant propaganda against the Armed Forces primarily on FINANCIAL grounds. Every year before the presentation of budget, discussion starts in newspapers calling for reduction of military expenditure. While need for increasing budgetary provisions for education and health is acknowledged, such increase cannot be made at the cost of Armed forces. Appropriate military expenditure is necessary if country's existence and sovereignty are to be safeguarded. Everything else comes after the safety of the state.

Let us now have a look at comparisons between defence expenditures of India and Bangladesh published in Daily Sangram. "India's population size and economy are both seven and a half times larger than that of Bangladesh. But her defence spending is 23 times higher than that of Bangladesh. It is, therefore, a matter of simple arithmetic that our military expenditure should be increased three-fold to match the Indian expenditure. The defence expenditute of Bangladesh in 1995-96 was increased only 2.5% over the previous year. India in contrast increased her defence expenditure by 18% in 1995—:96 over the last year's expenditure"(Daily Sangram : June. 23, 1995).

India is one of the poorest countries of the world. 30% of Indian people live in slums or bastees.50% dwellers of Bombay, 45% of Delhi and 65% of Calcutta live in the slum areas. 90% of Indian children live in malnutrition. 50% of Indian youth suffer from energy deficiency due to malnutrition. 216 million Indians do not have access to clean drinking water. 55 million children in

India are engaged in child labour. 75% of the primary school going children cannot afford to attend primary school due to extreme poverty. A World Bank report says "By the year 2000 A. D. the largest number of illiterates will live in India".

This poor country (India) has been building her military might and creating tension, rivalry and scare in the region. But RAW's stooges in Bangladesh are totally mum about it. If India agrees to cut her military expenditure, other regional countries will be too willing to reduce their military expenditure. Our defence forces are the protector and guardian of our independence. Despite RAW's. efforts, the Armed Forces are possibly the only institution in Bangladesh which have still remained relatively free of RAW's influence. But it does not mean that RAW has given up her designs against our Armed Forces. RAW has organised an elaborate espionage apparatus through its agents who are relentlessly trying to collect classified information and documents about our national defence. In the past many such RAW agents were arrested by police and our intelligence agencies. Here are a few examples. "Some RAW agents who were engaged in collecting maps and other secret documents of restricted area at Rajshahi have been arrested at Rajshahi. They include Sanjib Roy, Gaynath MandaI, Balram Mehata, Rampada Sardar and Ayazuddin. Most of them were hired to work for RAW man in the Indian Deputy High Commission at Rajshahi"(Daily Sangram : Decemebr 26, 1992).

"The police has arrested 5 persons for smuggling of valuabel state documents. Indian intelligence Agency RAW is believed to be involved"(Daily Sangram : December 11, 1992)."The law enforcing agencies have detained 10 persons including a former cabinet member of the deposed President Ershad in connection with a spy ring allegedly run by RAW. The agencies are looking for,a number of other persons for their involvement in the smuggling of confidential documents of the country"(Daily New Nation: August 15, 1993)."A former Indian diplomat, M.r.GR Makashi was expelled from Bangladesh on June 6, 1993. Mr. Makashi, who earlier was First Secretary in Indian High Commission in Dhaka was reportedly expelled on the charge of espionage. and anti-state activities. A local newspaper reported that Mr. Makashi had returned to Dhaka after resigning from service. He set up a garment factory in Dhaka's posh Dhanmondi residential area in partnership with others. He was being kept under watch by police due to his suspicious activities"(The Hindu International: June 19, 1993).

Indian High Commission at Dhaka and Deputy High Commissions at Rajshahi and Chittagong serve as hubs for espionage activities against Bangladesh Armed Forces. Besides, large scale trans-border espionage is also carried out by agents of Indian Military Intelligence, Border Security Force Intelligence, Intelligence Bureau and. notorious RAW as well. RAW's evil designs against Bangladesh Armed Forces are still continuing. We pray that this vital institution remains safe from RAW's menacing onslaught.

Campus Violence

The force that can effectively foil Indian hegemonistic designs is our student community. RAW has applied maximum efforts to extend its influence in our campuses to materialize her master plan. RAW is working on a long term programme directed to misguide and mislead student community and to spoil the congenial academic atmosphere in our educational institutions. Indians had been working in our educational institutions since the Pakistan period. Indian intelligence agencies had succeeded in influencing a group of student leaders during the student's movement of 1962.

This influence was further expanded during the 6-point movement. After creation of RAW in Septemebr 1968, India's influence on the student leaders increased substantially. As a result student-led mass movement of 1969 went, more or less, under the control of RAW influenced student leaders. It has now been confirmed by researchers, writers and journalists at home and abroad, including Indians that RAW had played pioneer role in creation of 'Joy Bangla Bahini'in 1970 and 'Mujib Bahini' in 1971.During the Pakistan days Indian secret agencies had kept their activities limited to influencing only the top student leaders. But after creation of Bangladesh RAW extended its canvas and now even middle and lower ranking student leaders are directly controlled and handled by RAW.Initially RAW spread its tentacles among the students and teachers of Dhaka University.. However, gradually they spread their operations to other universities and colleges including medical, engineering and other professional colleges all over the country.

RAW has three designs to achieve by paralyzing our educational institutions :

a. No resistance should develop against India and her cultural aggression, economic exploitation and political subjugation of Bangladesh. RAW has attained almost full success in this regard. During the last 24 years our educational institutions have played almost no role in protests against the aggressive designs of India. RAW's propaganda has made them a desperate, confused and misled force. Instead of resisting evil designs of India, they are engaged in negative politics, anti-social and even anti-national activities. Many students have started practicing Hindu culture in the name of Bengali culture.

b. India, through her agents, has constantly been creating clashes and conflicts in the campuses which have destroyed academic environments in Bangladesh. Due to ever increasing violence many students have started preferring to go to India for higher education. India thus not only earns a lot of foreign exchange but also is in a position to brainwash these students. When these students will return to Bangladesh after finishing their studies to take up responsible positions in Bangladesh, they will be more likely to have soft corner for India and thus will directly or indirectly serve her cause.

c. RAW has cultivated a strong pro-India pressure group among the students. Due to fear of pro-India students lobby the government can neither take any action against India's hegemonies designs nor even adopt an independent foreign policy. RAW has used a portion of our student community to put pressure on all governments from Mujib to Khaleda Zia by inciting the students to create

> disturbances and law and order problems. Commenting about the state of affairs of Dhaka University, The Daily Meillat wrote, "The Dhaka University, once known as Oxford of the East, had given birth to all the movements in this region since the British period. Many national movements were born in Dhaka University. RAW knows well that the anti-India movement, if any, would one day start from Dhaka University.

This is why a large section of the teachers and student leaders of the University have been engaged by RAW to keep 'development of any anti-India movement' under check. Many of them do not feel ashamed of openly acting against national interest. The University has become. a safe sanctuary for the terrorists and the murderers. Some RAW activists among the students and the teachers have become so bold that they have virtually made the whole University a hostage to their evil deeds. The University seems to have retreated from the main stream of our national hopes and aspirations. The terrorism which started from the Dhaka University campus has now spread over to other campuses as well.

The atmosphere for learning and teaching has withered away. That is why the college and university campuses are full of clashes, infighting, chasing, counter-chasing, demonstrations, gherao., gun-fights, arrests, injuries, killing, strikes, deferment of examination and even beating of teachers by students"(Daily Mellat : August, 25, 1995).There is not a single college or university in the country form Tetulia to Teknaf where clashes among the students do not occur. Many of these clashes are. not even reported in the media, thus the true statistics in this regard are difficult to compile.

However, following statistics published in The Daily Sangram detailing an account of clashes of four years (from January 1, 1991 to Febuary 15, 1995) provide some insight about state of affairs:

Clashes have occurred not only in the major universities of Dhaka, Rajshahi, Chittagong, Jahangirnager and Islamic University but also at far flung college of the island of Hatiya and at the mountainous region of Khagrachari. During this period due to Occurrence of clashes, the~nstitutions were closed down unscheduled for 250 times"(Daily Sangram : February 18, 1995).The Daily Ittefaq reported, "Including the death of Students League activist Joydip Dutta Chowdhury on January 21, 1995, 54 persons have been killed in Dhaka University premises since the inception of Bangladesh in 1971. Among them 20 were outsiders. Among them only two persons were killed in police firing, the rest were killed in clashes between rival students"(Ittefaq : February 28, 1995)."The students are killed in campuses due to internal feuds, power rivalry or personal quaI1rels. Enquiry committees are formed to investigate into these killiIgs.But none of the reports are ever made public for reasons unknown. Nobody has heard about any formal trial of the killers(Daily AI Mujadded : March 7, 1995).

According to Daily Janakantha,"In the Rajshahi University campus 23 perosons were killed in the last 23 years. Five of them were outsiders. The figure of injured is 1000. Nearly one thousand rooms were burnt down. The estimated value of loss on account of these clashes was Taka ten crores. The University remained closed for thousands of days during the last 23 years"(Daily Janakantha : February? 14, 1995)."Some institutions were closed down again and again. The Agriculture University Mymensingh was closed down unscheduled for ten times. In many cases when an institution opened after a long closure, it had to be closed again following bloody clashes"(Daily Meillat : August 25, 1995).

The RAW controlled terrorists continue their evil acts unabated in educational institutions due to patronage and support provided to them by influential RAW henchmen form within administration, businessmen, politicians etc. Some of our educational institutions have turned into mini cantonments through illegal import of arms form across the border. Commenting on the situation, The Daily Inqilab reported, "At present over 200 armd men are staying in the Dhaka University residential halls. Half of them are not students. These armed outsiders also reside in the halls. The statistics of the armed cadres staying in different halls are: Zahurul Haq Hall-30, Jagan Nath Hall-35, Zia Hall-20, Mujib Hall-35, Surja Sen Hall-30, Shahidullah Hall-25 and Fazlul Haq Hall-25. Among the arms held by these terrorists are cut-rifles, revolvers, sten guns, S.M.Gs, pistols etc. These terrorists are connected with influential persons who readily supply them with more sophisticated arms like sub machine guns, sten guns, mortars, SLRs, British carbines etc. whenever required".

The Daily Meillat published following report based on information obtained from intelligence and police sources, "The Dhaka University has become a hostage in the hands of 2000 armed cadres. Many of them permanently reside in different halls of the University. Most of them are outsiders. Among them there are 200 professional killers"(Daily Meillat : August 25, 1995).

RAW constantly remains on the look out to exploit issues for creating disturbances. When there is no issue, RAW uses its stooges to make a non-issue into an issue. RAW even creates dissentions in the same party,organisation. or institution. A big student organisation has been kept divided into Faridpur and Barisal groups. In the Jahangirnagar University campus, the students have been kept divided over locals and non-locals rivalry. At the Agriculture University campus, occurrence of fights between two groups—Dhaka and Mymensingh is common phenomenon. Besides, there are also occurrences of cutting tendon, veins, cutting off or breaking hands and feet, pulling out eyes, disabling or killing of rival students in various educational institutions.

"According to statistics available from Manabadhikar Samannaya Parishad, a Human Rights Organisation, some 30 educational institutions were closed amidst frequent clashes between activists of student organisations from January 1 to July 31 this year (1995). At least 16 persons were killed and about 8 hundred others injured in clashes. During this period 150 incidents of violence took place across the country"(Daily Morning Sun: August 15, 1995).

RAW engineered lawlessness in campuses has ruined educational environment. The situation has been further worsened due to dubious role-played 'by RAW influenced teachers. The institutions remain closed due to teachers' involvement in politically motivated agitation as well. During such closure, these teachers work with some NGOs for extra earning. Consequently the standard of education has fallen. As a result certificates awarded by Bangladeshi institutions have lost their value abroad. Commenting on the situation The Daily Janakantha wrote, "the guardians and parents being afraid of the falling standards of education and the session jams, prefer to send their wards to India for higer education(Daily Janakantha : February 20, 1995).

Over one hundred organisations have come up during the last few years that arrange admission of Bangladeshi students in Indian educational institutions. Advertisements are published in Bangladeshi dailies inviting students to seek admission in India. II Of late it has become fashionable for sons and daughters of bureaucrats, businessmen and politicians to study abroad particularly in India. They are going to India not only for higher technical education but also for subjects like fine arts, dance, music,,fashion, hair cutting, hair styling etc. It seems that the firms engaged as consultants for arranging admission of Bangladeshi students in Indian institutions have waged a crusade to attract as many students as possible. Incredibly even children of class I to V are also being sent to India for education.

According to conservative estimates nearly 100,000 students from Bangladesh are studying in Indian institutions. If we assume that average expenses per student per month are Taka 5000/-, the total expenditure comes to Taka 50 crore monthly and Taka 600 crore annually. These students are likely to be influenced by Indian culture, system and traditions. Their love for Bangladeshi literature, culture, history and tradition is likely to wane. Thus even of RAW does not directly recruit them, the Indian curricula would shape them as Indians at least in out 100k..It is also likely that a section of them would develop indifferent attitude towards our religious and cultural distinctiveness. These people are likely to be RAW's future lobby to demolish our independent identity and to work for merger with India.

It may be recalled that for Sikkim's merger with India the politicians and bureaucrats educated in India had played a vital role. Similarly some top leaders of Nepali a political party who had studied in India acted to make Nepal subservient to India even at the cost of national interests. Soon after the party rose to power, it rescinded arms purchase from China, adopted Indian diotated foreign policy and even abandoned King Birendra's initiatives for making Nepal a zone of peace (which was supported by 116 of the 159 members of the United Nations). the party also accepted unequal terms in case of the hydroelectric projects. All these things happened because minds of its leaders had been influenced by Indians while studying in India. Their education in Indian institutions had made them pro-India in their outlook and they became tools to serve for Indian interests.

RAW has also been using our student-folk for working against Islamic traditions and identity. They have also recruited some teachers for the purpose. These teachers give statements in newspapers and make speeches in seminars etc. in favour of secularism. They criticize Muslim traditions and identity. Due to lobbying by such secular minded teachers, the educational curricula of our universities have been based on secularism. Thus nobody was surprised when a teacher of History Department of Dhaka University recently demanded. for remember of Islamic history and culture from the syllabus.

After independence of Bangladesh Quranic verse 'Iqra Bisme nabbikallazi Khalaq' was dropped from the emblem of the Dhaka university. Similarly the word' Muslim' was dropped from. Salimullah Muslim Hall and Fazlul Haq Muslim Hall. Although Islam has now been recognised as state religion in the constitution but no step' has been taken by University authorities to re-introduce Quranic verse in the University emblem and the word' Muslim' in the name of the said two halls. When on June 8, 1994, in the annual sen,ate meeting of Dhaka university, Dr. Quamrul Ahsan Chowdury, the provost of Salimullah Hall proposed to reintroduce the word 'Muslim' alongwith the names of those two residential halls and the Quranic verse in the University emblem, the RAW influenced teachers and students rejected the proposal. Dt;. Qamarul Ahsan was labelled as communal. His office was attacked twice in succession by RAW terrorists-cum-students. They also threatened him for life. The University authorities took no steps to punish those terrorists.

RAW thus succeeded in its design to stop reintroduction of word Muslim and Quranic verse. Irony is bat it was accomplished through so called Muslim students.'The Bangladesh Times' reported that Hindu-Christian -Budha Oikya Parished had subinitted a memorandum to the VC Dhaka university stating that recitation of Bismillah and Quranic verses was contrary to the spirit of Liberation War. The news item says. 'The use of Bismillah in the programme of Dhaka University Platinum Jubilee celebration and DUCSU reunion has created resentment in a sedion of students of the University. It was revealed in a memorandum of DU unit of Bangladesh Chhatra Juba OikyaPrishad. (BCJOP), a sister organisaiton of Hindu- Budha-Ghristian Oikya Parishad, which was submitted to the VC. It was claimed in the memorandum that 'Bismillah' is contrary to the spirit of the Liberation War.

The memorandum was signed by more than 70 students (all belonged to Jagannath Hall). It was stated in their memorandum that in the original constitution of 1972 the pledges of the martyrs had been reflected, although these pledges have been stamped out through the cruel dissection of the constitution. The inclusion of Bismipah and reading of Qui-an fn the reunion programme of DU Platinum Jubilee celebration has struck a blow to the non-communal spirit of the University which will encourage to destroy the values of the Liberation War"(Bangladesh Times. September 24,. 1995).RAW is now pursuing an agenda to convert Dhaka University campus into a museum of statues. The aim is to gradually spread the statue culture all over the country.

New statues are constantly being erected on flimsy pretexts and baseless grounds. Many statues have already been erected in the campus. The plans are to erect many more statues of Hindu personalities by falsely associating them with the University It is well known that Hindus including Rabindra Nath Tagore had opposed the establishment of Dhaka University. When the University was finally established despite their opposition, Hindus sarcastically called it as Makkah University: The tragedy is that Dhaka University has now become main centre for spreading Hindu culture proposed by Rabindra Nath Tagore. If the process of erecting statues is continued, a day will come when visitors will find it difficult to believe that Dhaka University belongs to a Muslim country. It is to be remembered that Islam prohibits erection of statues.The Daily Meillat reported in this regard. "Keeping in view a particular objective, a neighoburing country has been pursuing the process of erection of statues in the campus of Dhaka University. In the span of ten years from now RAW plans to erect 200 statues in the campus alone"(Daily Meillat : May 27, 1995).

The Dhaka University which was established for the uplift of the down-trodden Muslims of this region preservation of their rights and identity, has now become' a tool in the hands of enemies of Muslim history and traditions. No one knows when Dhaka University will be free from the RAW's influence. The plight of other educational institution is not any better either. One can only pray to Almighty Allah to save our students from the clutches of RAW.

RAW AND OUR RULERS

Sheikh Mujibur Rahman, the founding father of Bangladesh and its first Prime Minister is regarded as a RAW agent by many analysts. However, the contention is not supported by many others. They argue that Mujib had elicited Indian nelp to liberate Bangladesh. India, on the other hand, used him to disintegrate its perpetual enemy Pakistan.

Their support was not due to any love for Muslim Bengalees. However; Sheikh Mujib utterly failed to foresee Indian game. It was only after creation of Bangladesh that Sheikh Mujip realized true Indian designs.

Indians expected that due to their help during liberation war Bangladesh will ever remain grateful and indebted. Particularly after signing of Friendship Treaty in 1972, Indians thought that Bangladesh will remain subservient and dependent on her. However, when the new country occasionally started conducting as a sovereign state, it was not liked by Indians.

RAW's apprehensions about Mujib grew due to the following reasons:

a. RAW feared that Mujib's popularity, towering personality and his ability to mobilize masses could pose a threat to India's overall objectives in Bangladesh.
b. Indians were annoyed when Mujib asked Indian Army to leave Bangladesh.

c. Indians were further irritated when Mujib ordered for return of Indian civil administrators who were sent to Bangladesh to take over the key positions after the fall of Dhaka on December 16, 1971.

d. Mujib's decision to attend summit at Lahore was not liked by Indians. India wanted that Bangladesh should remain a secular and a client state. They did not want it to develop relations with other states, especially with the Muslim countries.

In order to ensure that Indian interests are fully looked after, RAW arranged inclusion of its hard core agents in Mujib's cabinet as well as on higher positions in Awami League. These agents made all efforts to make Bangladesh a subservient state of India. This alienated the people from the Government. Yet RAW was not contented. Following 'divide and rule' policy, RAW managed divisions in Mujib's political followers soon after emergence of Bangladesh. Such is the ruthlessness of RAW that it did not hesitate from plotting against its friend and ally to further their own interests. Mujib's opponents later floated a political party which openly. preached and propagated severe bitterness and discontent against Mujib and his Government. Using it-RAW introduced in the country, a neW trend of political violence through hartals (general strike), blockades, sit-ins, bombings, arson and secret killing to halt the process of rebuilding the war devastated country. In the name of Gano Bahini (People's Army) an armed terrorist group was also formed by Mujib's opponents. In ensuing fights more than 10 thousand Gono Bahini cadres were killed (openly and secretly) by the then Mujib Government. Many of them included real patriots who had joined Gano Bahini with good intention of saving Bangladesh from Indian influence.

RAW used its stooges to forge Bangladeshi notes, arson jute godowns, created panic here and there by bombings, smuggled Bangladeshi goods and assets to India, organised insurgent groups like Shanti Bahini to disintegrate Bangladesh. Above all, RAW indirectly instigated Sheikh Mujib to form BAKSAL and one party rule which defamed him in home and abroad as a dictator and killer of democracy. RAW used Rakhi Bahini (a para military force created under Indian pressure which was virtually controlled by RAW) to eliminate nationalist forces.

Mujib did try to restrict RAW's influence but RAW's agents had infested his Government and party to such an extent that he virtually became helpless accomplice in RAW's game. Moreover his gratitude to Indians curtailed his will and ability to check RAW's activities. This led to sharp erosion of his popularity and mass support. Mujib alongwith his family members was assassinated on 15 August, 1975 in a military coup. After Mujib's assassination, Khondakar Moshtaq Ahmed took over the power. He announced to follow nationalist and Pro-Islamic policies. However, he was soon ousted in a counter coup master minded by RAW through Khaled Mosharraf.

Later Ziaur Rahman became the President of Bangladesh. He followed pro-nationalist policies and resisted Indian hegemonism. Zia's independent and Pro-Muslim foreign policy and above all, his obstinate stand on Talpatti Island made him the eye sore of India. Therefore, RAW turned against him. They instigated series of military coups to oust him. Ultimately RAW succeeded in killing him during 1981 after 18 aborted coups. The role of RAW in assassination of Zia is well documented in author's booklet titled, 'The Secrets Behind the killing of Zia'.

A noted Indian parliamentarian Mr. Subramaniam. Swamy,provided important insight about RAW's plan to assassinate Ziaur Rahman. He stated in an interview given to the weekly magazine 'Sunday', 'RAW had plotted the assassination of President Ziaur Rahman with approval from Prime Minister Indira Gandhi. Mr. R. N Kao, the RAW Chief and Shankaran Nair, a senior official of RAW had plotted to kill General Ziaur Rahman. The scheme was already at an advanced stage with Ganghi's approval. But the Congress Government fell and Mr. Morarji Desai became Prime Minister. Mr. Desai was appalled to learn about assassination plan. He called halt to the murder plot. The RAW warned the Prime Minister that it was too late to back out now and that many RAW assets would be endangered if the plan was aborted at this stage. However, Desai remained adamant and finally RAW abandoned the assassination'. Mr. Subramaniam Swamy commented, 'Zia continued to rule Bangladesh for many more years. He was assassinated after Indira Gandhi returned to power but India said that she was not involved'.

After Zia, Justice Abdul Sattar became the President of the country. He vowed to continue Zia's policies. However, soon his Government was overthrown and the then COAS Lieutenant General Hussain Mohammad Ershad assumed power on March 24, 1982 in a bloodless military coup. Ershad had attended an army course in India during 1974.

There are stories that he was picked up and cultivated by RAW during the said course. He was gradually nursed and supported to rise to the rank of COAS. It is widely believed that General Ershad captured power with the support of RAW. Reportedly Ershad had sent a special envoy to inform Mrs. Indira Gandhi about the planned military coupe to seek her blessings. Mrs. Gandhi gave her blessings on the assurance that Ershad would hold elections at the earliest and help in bringing into power a 'particular' political party.

However, later Ershad, developed his own ambitions. He started following independent foreign policy. He declared Islam as state religion. These moves annoyed RAW, who unleashed series of movements to topple him. Hartals, bandhs, blockades etc. which outwardly meant to fall his Government, also weakened the economy of Bangladesh and discouraged foreign. These factors impeded country's progress and development. During Ershad regime RAW regained ground lost during Zia era. He patronized Indian lobby and allowed RAW to conduct its activities virtually at will. The appeasement policy followed by Ershad enabled RAW to dig its roots deep in administrative, cultural and political echelons of Bangladesh.

Ershad's belief was that by giving maximum concessions to the Indians, he could win their support, however, proved misplaced. He failed to understand that Indians did not want merely concessions, but total subservience, which perhaps was a little. too much for him. Ershad was forced to resign on December 6, 1990 in the wake of popular uprising and wide spread public agitation. After Ershad, Bangladesh Nationalist Party (BNP), led by Begum Khaleda Zia came to power by winning elections on anti India and pro-nationalist stance. People hoped that BNP would follow bold and courageous policies. However, the performance of BNP Government so far (1991-95) is indicative of the following:

a. BNP has been trying to outdo Awami League in the ideological and other fields which were once considered the latter's domain.

b. The basis of foreign policy of BNP Government has been appeasement of India.

c. RAw's activities and operations in Bangladesh have continued unabated. The RAW agents and stooges in higher bureaucracy and the Government continue to thrive. On the other hand, the nationalist and anti- Indian officials have been targeted and side lined. A glaring example of the same is tpe transfer of Mr. Assafudollah, the then Secretary of Agriculture and Water Resources Ministry. He was sent out of the Ministry due to his uncompromising and obstinate stand on Farrakka issue. On the contrary, Mr. Farooq Sobhan the Foreign Secretary, who reportedly has been advocating for grant of transit rights to the Indians still continues to occupy the coveted appointment, despite serious allegations leveled against him by the nationalist press.

d. Under the cover of free economy Bangladesh has been over-flooded with Indian goods,which has caused grave harm to country's industrial development. Many of the existing Bangladeshi mills and factories are on the verge of closure due to unplanned in-flow of Indian goods. The New Nation, an English daily wrote on 29 August 1995, under caption 'India seek cooperation from Bangladesh'. 'Trade between the two countries now stands at 60 crore dollars compared to that of seven crore dollars five years ago'. It may be mentioned that balance of trade is favourable to India by 97% to 3%.

e. A large number of newspapers', magazines, publishing houses, business firms etc. have been floated by RAW under dummy ownership. These are relentlessly engaged in sabotaging the country's ideology, independence and sovereignty.

f. Polarization on account of Freedom Fighters versus Non Freedom Fighters received a new surge.

g. Educational institutions experienced highest frequency of violence and terrorist activities. This situation actually was arranged by RAW hirelings and the process spread throughout the country like contagious disease. Because of violence, murders and session congestion in the educational institutions, large number of students went to India to study in Indian universities and colleges. As a result a huge amount of foreign exchange is draining down to India. Moreover, it has provided good opportunity to RAW to brain wash our younger generation during their stay in India.

h. No action has been taken against RAW stooges who work or speak openly against our independence and sovereignty. Our Government failed to take appropriate legal action against Taslima Nasreen, Ahmed Sharif, Hassan Imam and leaders of the Hindu-Bhudha-Christian Okkayea Parishad for speaking and writing against our country and religion. On February 28, 1992, an Indian journalist and Editor of Dainik Ajkal, Mr. Myaram Surgeon in a seminar held in the National Museum Auditorium Dhaka, urged Bangladesh to merge with India. But no protest was made instantly or later, though the Prime Minister of Bangladesh Begum Khaleda Zia, the Leader of the Opposition in Parliament Sheikh Hasina and other top-level leaders were also present there.

It is apparent that most of our rulers have been following the course of least resistance against RAW's onslaught. Some even resorted to appeasement to get RAW's blessings. There is a fallacious dictum in Bangladesh: 'To rise to power you need to be anti-India but to survive in power you need to appease India'. Perhaps due to belief that no Government in Bangladesh can survive if it invites wrath of RAW, our rulers do not dare to challenge and bridle RAW. However, I vehemently oppose this supposition. A

Government which boldly, courageously and righteously pursues national interests will become so popular that no body will be able to shake its foundation.

Here I also want to dispel fears propounded by some cowards and pessimists that since Bangladesh is surrounded by India on all sides, therefore, it is defenseless. They argue that due to its size and power, India is in a position to destabilize Bangladesh at will. I want to draw their attention towards history which proves that it. is not the size alone which determines a country's defensibility. After all, small countries like Cuba and Taiwan have existed despite presence of mighty hostile countries around them. Similarly examples of Afghanistan and Vietnam can be given who being much smaller and poorer, still defeated super powers like Soviet Union and the United States respectively.

Thus the true guarantee of a country's independence and sovereignty. are will and determination of its people. The size, affluence and resources come later. The Bangladeshi people are indeed very patriotic and motivated. History has proved that they are willing to Fender any sacrifice for their motherland. Hence our rulers should rely on our nation's spirit and motivation for upholding and safeguarding country's integrity and independence. They should understand that unnecessary appeasement of foreign powers will not earn them any good. Our adversary is too wicked and has too ambitious an agenda. Mere appeasement does not satisfy his greed and hunger, hence it will prove futile. Those who propound the 'appeasement theory' are either myopic or fifth columnists.

Let it also not be misunderstood that I am advocating anti-India hysteria. Certainly that" is not my aim or objective. What I am trying to put across is' that we must develop the will and conviction to preserve our independence and sovereignty. We must not allow anyone a free hand to subvert our ideology, independence and national interests. We should assert ourselves whenever required for claiming our just rights. That is the only course for an honourable existence. Let us hope that our rulers and leaders will realize this' simple fact in their own as well as in country's interest.

RAW'S INFLUENCE IN HIGHER ECHELONS

RAW has carefully nurtured a strong pro-Indian lobby amongst top political leaders and highly placed Government officials for promoting Indian interests in Bangladesh. Many of them are wooed through rewards in cash or kind. The others tow Indian line due to their ideological leanings. RAW's amazing success in penetrating in our top echelons is a cause of grave concern. Indeed vital national interests will continue to be jeopardized if the situation is not remedied. Here are some. examples of top level betrayals. The Weekly Bikram of Dhaka, in its issue of 29th November, 1993, published an interview of the then Assistant Secretary of the Ministry of Shipping and Water Transport. A relevant part of the same is reproduced below:

'In 1972 the Ministry of Shipping had no sea-going vessel. The Prime Minister's Secretariat directed this Ministry to procure a ship of its own. Soon we processed the file and sent that to the Ministry. The Finance Ministry in their note told us to look into the matter whether India could supply us such a ship, otherwise India's approval would be required if we wanted to procure it from some other country. General Osmani was in charge of Shipping Ministry while Mr. Tajuddin was Finance Minister. After receiving the note of/the Finance Ministry(Daily lnqilab : December 16, 1994)took the file to the Prime Minister's office. Sheikh Mujib, the then Prime Minister got furious after seeing. the note. He summoned Mr. Tajuddin to his office. No sooner had Mr. Tajuddin entered the Prime Minister's office, then he burst at top of his voice and said, 'What else have you, Tajuddin conceded to India?'

The Daily Meillat in its editorial titled 'Allocation of money for the Ganges Barrage' exposed the extent of RAW's influence on our highly placed persons. It wrote :

"The Agriculture Ministry has stated that the Government does not consider the Ganges Barrage project as an alternative to the Farakka Barrage. This amounts to denying the utility of the Ganges Barrage. It is not known whether the comment was based upon facts and findings by the water experts or it was yet another attempt to please the Indians. The feasibility of Ganges Barrage has been well assessed by many experts who had recommended implementation of the project. Based upon the evaluation of experts, late President Ziaur Rahman had started the project and he himself laid the foundation stone of the Barrage.

India and her local agents have been opposing the project for ulterior reasons. Indian agents in Bangladesh's higher echelons have been delaying the implementation of the project on one pretext or the other(Daily Meillat : May 24, 1995).The Daily Janakanta reported how some of our senior administrators and bureaucrats work for India, relinquishing country's interest. The report alleged that although the Government had prohibited import of granulated SSP fertilizer from India, yet the Ministry of Agriculture issued a circular on April 17, 1995 to import Indian fertilizer. It is further stated by the newspaper that the Agriculture Ministry in a meeting on October 24, 1994 had decided to stop production of SSP fertilizer in Chittagong TSP Complex and instead decided for production of 100% TSP at the complex.

Bangladesh is self sufficient in TSP production -and can even export some surplus. But production of SSP is lesser than the domestic requirement. SSP and TSP can be produced in the Chittagong factory at the ratio of 60% SSP and 40% TSP. Bangladeshi SSP is considered superior to Indian SSP. The use of SSP can increase agricultural yield by 10%. Even being aware- of all these facts, - the Agriculture Ministry asked the Industries Ministry to stop production of SSP. The newspaper alleged that this was done to import larger quantities from India to make up the shortage at later stage. Indian SSP which looks similar to Bangladeshi TSP, can be sold at higher price (price of TSP is higher than SSP). The Daily comments that the said lobby is working actively to make Bangladesh a market for Indian goods. They do not care if in the process country's economy and industry are totally destroyed(Daily Janakantha : June 20, 1995).

The Daily Inqilab reported about appointment of an Indian in Bangladesh T & T Board. The Daily stated that a Director of the Delhi Communication Centre of India got the appointment at the T&T Board without prior approval of the concerned ministry. The appointment was made disregarding the time frame that was given to the. International Telecom Union for sending their nomination for the post. The Board had not even cared to look for an expert from another country. According to the details T&T Board had approached International Telecom Union to nominate an expert for Data Exchange Centre built by it. The International Telecom Union sent the name of an, Indian expert. The concerned Ministry did not approve the proposal and asked for a panel of three names for final selection. However, disregarding the Ministry's instructions and before getting the desired response from Telecom Union, T&T Board decided to appoint the Indian expert who was previously rejected by the Ministry. The Indian expert reached Dhaka and assumed his new assignment.

When asked, the Ministry of Post and Telecommunications expressed its ignorance about the said appointment (Daily Inqilab : December 7, 1994).After publication of above news, the T&T board sent a clarification to the Daily Inqilab justifying their decision. But the reply was unconvincing. They could not give any justification about why an Indian was appointed on such an important and sensitive position.

The transport sector of Bangladesh is dominated by Indian vehicles. The Suzon Tempo of Bangladesh has lost its market due to Indian Bajaj and API auto rickshaws, which have been given excessive concessions. Similarly, Indian locomotives have been preferred over better quality locomotives from Britain, Germany and Canada. This was made possible by winks from right quarters despite the fact that the Railway Tender Committee had rejected the offer of Indian Locomotives Company. The Tender Committee had concluded that maintenance cost of Indian locomotives will be, seven times higher(Daily Inqilab : November 23, 1994).

The main objective of importing Indian locomotive engines is to make Bangladesh Railways dependent upon India and also infiltrate Raw agents in Railways in the guise of experts and engineers who will be coming from India to maintain and repair the locomotives. Let us see what The Daily Inqilab Comments: 'India is going to succeed in grabbing Bangladesh Railways beca1,LS,e of the intrigues of a section of vested interests. India began assembling locomotive very recently and the industry' is still in initial experimental stage. The Railway's Tender Committee had mentioned in its report that Indian locomotive engines are totally unfit and unacceptable for Bangladesh and had recommended that Bangladesh should not purchase these engines under any circumstances. The committee also mentioned that yearly maintenance and' running cost of the current engines with Railways is one tenth of the total cost of the engine, whereas the said cost of the Indian locomotives will be more than seven tenth. Thus Railways will have to incur seven times more expenditure on their maintenance. Besides, the Indian engines are of lower quality and the engineers of Bangladesh are not experienced and habituated in managing and driving them. Hence Indian experts will have to be hired to supervise and maintain these engines. Despite such strong opinion of experts committee, the decision was made to purchase Indian locomotives(Daily inqilab : November 23, 1994).

The above case makes RAW's influence in the Ministry of Roads and Communication amply clear. As a result of the said arrangement Bangladesh Railways will be required to spend extra amount on maintenance. Besides, the Indian engineers will stay in Bangladesh for seven years to maintain and repair these engines costing additional 60 million US dollars. It is apprehended that RAW will be able to place its agents in 'Bangladesh" Railways in the guise of experts and technicians. Bangladesh is self sufficient in salt production.. There are huge stocks of better quality iodine salt in the country. But still some dishonest businessmen have been given permits to import salt from India. They import non iodized cheaper salt which is full of adulteration of sand, stone, glass particles etc. The Indian salt is also inferior in chloride content i. e. 55% as compared to 99% in Bangladeshi variety. Consumption of low quality salt causes various diseases, hence its import should not have been allowed.

On December' 11, 1994, a unit of Bangladesh Navy captured an Indian Trawler alongwith 15 Indian fishermen in Bangladesh waters, They also had Rs. 126853 in their possession. A case was registered in the Mongla Thana in this connection on December 12, 1994.But for mysterious reason the ministry of Home Affairs issued an order to the Magistrate of Bagherhat. Accordingly, the concerned Magistrate on December 26,1994, withdrew the case and ordered the authorities of the Bagherhat 'prison to release the Indian citizens alongwith their possessions. When contacted, the Home Ministry refused to explain why the case against the illegal infiltrators was withdrawn (Daily Meillat : December 30, 1994).

Bangladesh needs to import sugar from abroad to meet short fall in local sugar production. But the irony is that sugar imports from India far exceeds country's actualrequirements. As a result the Bangladeshi sugar remains unsold due to dumping of cheap Indian sugar.The Daily Al Mujadded reported on June 22, 1995, that 133, 200 tons of Bangladeshi sugar remained stock piled and unsold. The Daily Sangram reported on June 21, 1995 that Bangladeshi sugar worth taka 4000 million remained unsold. The daily.

also reported that although import of more sugar was no more allowed, yet a dealer opened an L. C. account with a bank for import of 200,000 tons of sugar primarily from India. In addition smuggling of sugar from India is a common practice. Low quality cheaper sugar imported from India has adversely affected the domestic sugar need to pay VAT at Taka. 2.42 per Kg,whereas sugar entering Bangladesh through smuggling need not to pay any tax or tariff.As a result Indian sugar is sold at a cheaper rate. Often it is found that Indian sugar contains adulterated items like urea, sand and even glass particles, but no remedy is available to redress the complaint.

Indian edible oil has also captured Bangladeshi market. Smuggled Indian edible oil is cheaper by Taka 18/- to 20 per Kg as no duty or VAT is levied on said edible oil. Faced with inequitable competition; oil refining mills in Bangladesh are being closed. Out of 60 mills, 40 have been already closed down. Production of oil has fallen from seven lakh tons to only three lakh tons. It is estimated that four takh tons of edible oil is smuggled into Bangladesh from India each year (Daily Sangram : June 21, 1995).

When the Government recently decided to import food grains to make up the shortage, some highly placed RAW agents were able to persuade the Government to import maximum food grains from India on the ground of its proximity. A little quantity of food grains were also imported from other countries including Pakistan, which was merely an eyewash. There are complaints that most of the Indian rice did not meet the conditions of sample of the tender. Indian rice was sub-standard, adulterated (with stone, gravel and grit), rotten and worm-eaten. Numerous reports were published in local newspapers ventilating the sub-standard and inedible position of the Indian rice. However, defying all criticisms the concerned authorities accepted the rice.

The role of the officials of the Ministry of Food in the case of 8 thousand metric ton's of cotton and unfit rice imported through Khalil & Co. is indeed revealing of their subservience to India. The said rice consignment reached Mongla Port on June 11, 1995 by an Iranian flag vessel.

The firm appointed to off load the rice refused to do so' as the rice was rotten and worm-eaten. Local food officials inspected the rice and informed the Ministry of Food in Dhaka. To handle the situation the Chief Marketing Manager of the Indian State trading Corporation, Mr. Tendon rushed to Bangladesh. This reversed the situation completely. The contractor that had refused to unload the rice was fined taka 11 lakhs and black listed on the instruction of concerned authorities in Dhaka. Another firm was assigned the responsibility to unload the rice at higher rates. The firm refilled the rice in new sacks inside the ship, though the rice was really rotten and worm-eaten(Daily Al Mujadded : June 19, and 25, : 1995).

Another step taken by RAW towards crippling the agriculture and economy of Bangladesh is the supply of poisonous pesticides to Bangladesh which are prohibited even in India itself. These include Thiodine, Hilbil, Bittil, Hexactol and Thionyl etc. These pesticides are also prohibited in Bangladesh but their smuggling from India is exceedingly rampant. Informed sources claim that RAW has arranged the production of these illegal and destructive insecticides in India for supplying these to Bangladesh through smugglers, who are i~ fact RAW agents. It is but natural that prohibited items cannot be produced by any private firm without direct instructions and protection of Government authorities in India.

Thus it can be easily inferred that their production and smuggling into Bangladesh is taking place with connivance of RAW. These insecticides are injurious not only for the crops but also for human beings. They are so poisonous that their destructive power cannot be destroyed even at 600 degree centigrade temperature. People who eat the vegetables on which these insecticides were used or applied may suffer from paralysis, dyspepsia, heart and liver problems etc. They also weaken different important internal organs of human body. Thiodine and some other insecticides reduce the fertility of the earth. Besides; those who eat the rice from the field where Thiodine was applied may suffer from dysentery, weakness and dyspepsia. Agricultural experts predict that within next 15 years the fertility of the land of Bangladesh would be reduced to an alarming extent. Observers believe RAW produces and send these insecticides to cripple the economy, ecology and human resources of Bangladesh. Insecticides are available on cheap rates which easily attract the illiterate and poor tillers of Bangladesh. For unknown reasons no firm step has yet been taken to stop the illegal trade of these poisons.

The Ministry of Foreign Affairs is another hot bed for RAW stooges. It is notorious for its pro-Indian stand on all issues, national as well as international. The influence of RAW on our Foreign Ministry 'was clearly-unveiled when our Foreign Secretary Mr. Farooq Sobhan himself took the initiative to propagate for grant of transit facility to India.The Daily Inqilab informed that Mr. Farooq Subhan had urged some journalists to create favourable public opinion for awarding transit facility to India. It is to be noted here that prior to his current assignment, Mr. Farooq Sobhan was the High Commissioner of Bangladeshi in New Delhi. Daily 'Inqilab further commented, 'It is clear that the Ministry of Foreign Affairs of Bangladesh has taken the responsibility of safeguarding the interests of India.

The recent statements and propagation of this Ministry to neutralize the displeasure of people against India due to her anti Muslim role, prove its pro-India bias(Daily Inqilab : July 29, 1995).At a premier institution concerned- with national defence, students frequently ask the guest speakers, 'Since Bangladesh does not have viable defence potential, therefore, would it not be feasible to come under security umbrella of India?' Apparently the question denotes mere academic probe, but its ramifications in terms of thinking and belief of those who are supposed to work and plan national defence are indeed much larger. RAW, through its continuing propaganda seems to be affecting even partiotic minds on the plea of inviolability of national defence.

RAW's involvement in internal politics of Bangladesh is deep rooted. Since the days of erstwhile East Pakistan RAW has been patronising and funding a particular group of political parties(The Gentleman, Bombay,November, 1985).RAW's interest in political activities remained alive even after establishment of Bangladesh. The RAW operatives based in Indian High Commission Dhaka, maintain regular contacts with Bangladeshi political leaders and uses them for furthering Indian interests.

A report about involvement of RAW with certain political party in Bangladesh, compiled by Mr Rajesh Joshi was published in Daily Indian Express. It was later re-produced in Dainik Bangla on 3 April 1992 and Daily Meillat on May 7, 1992. The report says, 'RAW had given 4.50 crore rupees to a political party of a neighbouring country (Bangladesh) for elections held on 27 February 1991. Mr B B Nandi (a high official of RAW) had given the money at Calcutta. But ultimately that political party failed to win the elections'.

There are reports that money was paid to some other smaller parties and individuals as well. There are also reports that RAW had offered to pay money to some other parties including pro Islamic and pro-nationalist parties. But they declined to accept the money. The point to be noted is that RAW tried to enlist support of diverse political groups who are ideologically opposed to each' other. It proves that RAW is working on the policy of having two types of clients in Bangladesh i.e. A team and B team. This is to ensure that irrespective of which political party is in power, their purposes should be properly served.

In pursuance of above policy, RAW has cultivated agents and supporters in almost all major political parties. These highly placed persons are provided material incentives in cash as well as in kind.

Grant of agencies of Indian companies, promoting business, helping in securing loans, arranging medical treatment in India for them and their families, granting scholarships to The children of favoured politicians are only some ways to dole out incentives to their agents and supporters.

A recent example of RAWs influence over the Government circles of Bangladesh was the proposal of giving the responsibility of preparing the voters lists and laminated identity cards to Indian companies. Sixty companies, both local as well as foreign, had submitted tenders for this work. But the concerned quarters selected only Indian companies to undertake this work of national importance. It may be' pointed out that necessary material, technology, organization and expert manpower were available in Bangladesh for doing the said work. A Bangladeshi firm, Swapta Sindhu Limited had offered to do the job using local technology and manpower within 100 days at a cost of 150 crore taka. But despite their competitive offer the work was assigned to Indian companies.

Had not the patriotic people of Bangladesh protested against this grave security lapse, RAW would have succeeded in its nefarious designs. Taking advantage of this opportunity RAW would have been able to enlist many Indian Hindus as voters in Bangladesh to achieve favourable election results. By such manipulation RAW may have succeeded in bringing to power a puppet Government in Bangladesh to serve their purpose. Such a Government could even be expected to go to the extent of proposing in the parliament to incorporate Bangladesh into the Indian Union. Another important benefit for the Indians would have been that once credentials of these Hindu Indians were established as citizens of Bangladesh, it would have been easier to push them back for permanent settlement in Bangladesh.

Even the Hindus who had left Bangladesh during 1947, after disposing off their movable and immovable properties and had been settled in India as permanent citizens, might be brought back to work as agents of RAW. A campaign had already started propagating that nearly one crore Hindus were forcibly compelled in the past to leave Bangladesh. An organization called Hindu-

Buddha-Christian. Oikya Parishad, which works as an offshoot of Indian BJP in Bangladesh, in their national conference held on May 4, 1995 demanded to bring back these one crore Hindus from India. Although after storm of protests from all over Bangladesh the award of said work to Indian companies' was cancelled, but an important point to be noted is that some of the opposition parties who otherwise routinely criticize every Government action did not utter a single word against this decision.

Is it not proof enough of RAW's influence on Government as well as on many political parties in the country? However., the redeeming feature is that people of Bangladesh stand solidly together to thwart Indian designs. Indeed that is the real guarantee for country's independence and sovereignty. The above are only some of the examples of RAW's influence in higher echelons of Bangladesh. I could go on narrating endless details of the same but the paucity of space restricts further narration. However, I hope that the readers would have grasped the magnitude of RAW's penetration.

THE TRANSIT ROUTEA

India has been demanding grant of transit route through Bangladesh for providing supplies to its seven north-eastern states. The proposal is that Bangladesh should provide road and rail transit between Calcutta, Agartala, Shillong and Gauhati via Dhaka. Besides, use of Bangladesh's inland water-ways for Indian steamers is also demanded. India also wants to use Chittagong Port for transportation of goods to Tripura and Mizoram states. To make these proposals profitable a toll payment is being promised. Such transit facility arrangements do exist in several parts of the world, therefore Indian proposal at the out set may appear innocent and innocuous. But the problem arises due to India's expansionist and evil designs against her smaller neighbour. The fear is that the use of-transit facility will not remain restricted for commercial purposes. Rather it will be used for subverting Bangladesh's sovereignty and economic well being.

It may be pointed out that the northeastern Indian region is not isolated or cut off from mainland. Already good communication links exist between India's mainland and the seven northeastern states. These channels are already being used not only for supplying consumer goods but also for movement of troops. Thus there is no dire necessity for a transit facility through Bangladesh. A dispassionate analysis suggests that these proposals should not be accepted by Bangladesh for the following reasons:

a. A wide trade gap in favour of India already exists due to Bangladesh's weaker economy, Indian machinations, unbridled smuggling and dubious economic policies by our Government. The grant of transit facility to India will facilitate smuggling, thus increasing the trade gap even further.

b. It will not be possible for our police and security forces to supervise Indian vehicles and commodities throughout the proposed transit routes. At best the checking may be possible at entry and exit points though even that will depend upon cooperation of the Indians. India may cooperate during checking of vehicles and commodities in the initial days, gradually she would backout. It should not be forgotten that India has already earned a name for breach of promises and wriggling out of her international commitments.

c. India may initially sign an apparently amicable transit agreement to make a beginning. But at a certain stage she may start demanding amendments in the terms and conditions of the original agreement. It will then be difficult for Bangladesh to get out of the net.

d. RAW will certainly use this facility for furthering, and pursuing its own agenda of subversion. It will become easier to smuggle arms and explosives throughout Bangladesh. Besides, it will facilitate movement of RAW's agents, terrorists and stooges to and from Bangladesh.

e. Owners of Indian buses and trucks will be greatly benefited by transit facility. On the contrary, Bangladeshi transport sector will suffer due to adverse competition.

f. Another fear is that the said transit facility will not remain restricted to commercial use only. It is almost certain that the Indians will use it for military purposes on the pretext of utmost necessity. The movement of troops and supplies of arms to the northeastern states will gradually become routine. This can be used by India for massing troops even inside Bangladesh.

g. The said facility will also be used by India for troop. movements against China. This will hamper. Bangladesh's relationship with a friendly country. "

h. In the name of ensuring security of Indian property and personnel, India may demand from Bangladesh for allowing her to station her troops inside Bangladesh. RAW may arrange feint attacks on Indian vehicles through its agents and thus create enough justification for India's demand to station her troops to protect her persons and property. In this garb RAW may get free hand to physically annihilate patriotic forces opposed to Indian expansionist design. Another danger is that RAW may even arrange mock attacks on Indian troops to 'prepare grounds for India to undertake offensive action against Bangladesh.

i. Eventually it will become impossible for Bangladesh to exercise her sovereign authority on the transit route. But once allowed it will be very difficult to close the same. India may argue that the transit route is vital for her economy and security and its closure will amount to hostile action. In such a situation India may even force war on Bangladesh and capture her, to protect her vital interests.

RAW has recently stepped up campaign to press for 'transit facility'. Its stooges are engaged in propagating the advantages Bangladesh would have if she accepts India's proposal for transit facility. Beside India is exerting relentless pressure on Bangladesh for grant of transit facility at least on experimental basis. Of late efforts are being made to' link the issue with Farakka Barrage and proposals far joint resolution of the two' issues are being put across. It has also been offered that as a reciprocal measure Bangladesh may be granted transit route to' Bhutan and Nepal. However, we must not fall prey to' any such tactics. We must not forget the character of Indian leaders. If Indian bus and truck services once start plying through Bangladesh, it will never be possible to' stop them under any circumstances. We remember how India withdrew water of the Ganges at Farakka on an experimental and temporary basis. Sheikh Mujib trusted the sweet promise of Indian leaders, in good faith. But that trial operation of Farakka still continues and perhaps it will never came to' an end.

It is obviously difficult to' stop anything if it once starts. Sa if Bangladesh once grants transit facility an experimental basis, it will be next to impassible to stop it given India's track record. It should be remembered that India may even apt far open hostility including declaring war against Bangladesh, if she considers that her vital interests are being endangered by our actions at same stage. On the other hand India can withdraw the reciprocal concessions given to us in exchange of transit facility on any flimsy ground and we would not be able to do anything about it. In the face of such possibilities we have no reason to be attracted to the sweet bait of Indian leadership. It will never be beneficial far Bangladesh to grant transit facility to India in return far same monetary benefits no matter what it amounts to. Why should we allow India to use our land in the name of transit facility when we are certain that it will impinge an our sovereignty and national interests?

ANTI-BANGLADESH PROPAGANDA ABROAD

RAW is continuously spreading false propaganda against Bangladesh specially, in Europe and United States of America. RAW's objectives in this regard are:

(1) TO defame Bangladesh.
(2) TO isolate Bangladesh from the community of nations.
(3) TO project Bangladesh as a fundamentalist country where religious minorities do not enjoy basic human rights.
(4) TO influence policies of Western World through misinformation as that their economic and political support to' Bangladesh decreases.

With a view to achieving above objectives RAW is pursuing vigorous anti-Bangladesh campaign in Europe and USA particularly far the last few years. In 1992 RAW engineered the setting up of Bangladesh-USA Human Rights Group based in the United States. The stated aim of the organization is to maintain human rights situation in Bangladesh. However, it has become a forum far spreading anti-Bangladesh propaganda. Its reports to the contrary tend to serve Indian interests.

Commenting about the said organization Weekly Dhaka Courier wrate, "Initially people welcomed its inception but it failed to conceal its ugly face and started doing what served India's interest better. Just after its inception it wrote a letter to the US Congressman Tom Lantos, Co-Chairman of the Congressional Human Rights committee bringing multiple charges of human rights violations on minorities in Bangladesh. This group appealed to the US Congress to stop American financial assistance to Bangladesh unless the religious minorities are properly treated.

The letter alleged that Bangladesh Government had reaffirmed Islamisation in Bangladesh and ignored the rights of minority communities. It alleged that Bangladesh violated UN Human Rights charter by infringing upon religious rights of Hindus, Buddhists and Christians. Copies of this letter were circulated among the high ranking members of the Congress and' the Senate (Dhaka Courier: February 5, 1993).

A 3-day international conference was held in London from October 18 to 20., 1991 on the status of minorities in Bangladesh. A number of so-called delegates were invited from different parts of Europe and The United States to represent Hindu, Buddhist and Christian communities of Bangladesh. However, in actual fact, most of the so-called delegates were either Indian citizens or Americans or Europeans of Indian origin. Only a few delegates belonged to Bangladesh. The speakers at the conference spoke about grossly exaggerated accounts of maltreatment, discrimination, atrocities and forceful exodus of the minorities in Bangladesh.

Dr. Sabyasachi Dastidar, a Professor of New York University who had migrated to India in the late 40.s from Barisal District of Bangladesh presented a paper at the conference on behalf of the US branch of the Hindu-Buddhist-Christian Council. The paper was later published in South Asia Forum, a quarterly brought out from New York to draw the attention of the United States Government and people towards fake charges against Bangladesh. During the later part of 1992, RAW arranged numerous seminars, meetings and demonstrations in New York highlighting the subjects related to mistreatment of Bangladeshi minorities. Guest speakers were invited from foreign countries in these seminars and meetings to use them in anti-Bangladesh campaign. Articles, features and newspaper reports about discrimination against minorities in Bangladesh published in pro-Indian newspapers in Dhaka were translated and distributed to different activists and organisations.

RAW has also been making concerted efforts to internationalise CHT problem. In 1992 RAW stooges arranged a conference in New York School for Social Research on tribal minorities in South Asia with speical reference to Bangladesh. One Bangladeshi speaker tried to draw a parallel between the 1971 war of liberation and the armed struggle of tribal people in the Chittagong Hill Tracts. RAW arranges posting of fake letters to the US Congressmen, Senators, human rights groups, United Nations affiliated agencies etc. about alleged human rights abuses in CHT.

Further it arranges publication of articles in the media to draw attention to the so-called atrocities of the Bangladesh Armed Forces on the tribal people of CHT. Priests and posters of churches are also approached in the United States to write letters to local Congressmen and put political pressure on the US Capital Hill on the issue. A lengthy memorandum was submitted to the UN Secretary General requesting him to send a UN team to Bangladesh for investigation of alleged war crimes. Copies of protest letters and leaflets were distributed in the streets of New York to create favourable public opinion in this regard.

RAW selected New York as the main centre for anti-Bangladesh campaign because: (1) It is the nerve centre of world politics and business where leaders, intellectuals and businessmen from all over the world get. together. (2) USA is now the only super power. Besides it is the 2nd largest donor country to Bangladesh and the largest buyer of Bangladeshi garments. Hence the effort to hamper Bangladesh-US warm relations. (3) To create an anti-Bangladesh public opinion so that USA does not put pressure on India to resolve its disputes with Bangladesh.

Influenced by these organised campaigns in the USA during the last couple of years, 17- US Congressmen in a joint letter to Prime Minister Khaleda Zia expresed their concern on the reported discrimination and mistreatment of religious minorities in Bangladesh. In the letter they said, "The establishment of Islam as the state religion continues to cause concern among Hindus, Buddhists and Christian minorities. We are concerned that these events are part of mistreatment of religious minorities. Among the Congressmen who signed this letter (from Democratic as well as Republican Party) are: Thomas Foglietta of Pensylvania, Gary Ackerman, Eliot Engle and Edolphus of New York, Henry Waxman and Nancy Pelosi of California, ehristopher Smith of New jersey and Bernard Sanders of Vermont.

The RAW stooges like Bidut Sarkar, Stephen Gomez, Shambal Chowdhury are being protrayed as leaders of the Hindu-Buddhist-Christian Council in some US newspapers. A Buddhist monk named Bimal Vhikhu has been going strong lobbying in Washington and the United Nations. Bimal Vhikhu had migrated to Calcutta in 1986 and permanently settled in India. CR Dutta, Bimal Vhikhu and other leading activists met several members of the House of Representatives and Senate including Senator Kennedy in Washington. It is learnt that RAW has been using Indian missions in Washington, New York and other. US cities for arranging interviews of dissident leaders with the leading Congressmen an,d Senators.

RAW has also been using various UN bodies for propagating against Bangladesh. In 1992 Centre for Human Rights based in Geneva arranged a conference on the International Year of the World's Indigenous People. RAW arranged inclusion of an Indian citizen Mr. Bimal Vhikhu in the list of the NGO speakers. He addressed the conference claiming to represent the tribal community of CHT. He castigated Bangladesh for all sorts of actions and excesses against tribal people of CHT.

RAW, has also been arranging visits to UN Human Rights Commission by Chakma leaders almost every year. After the demolition of Babri Mosque, Bangladeshi Muslims showed highest degree of restraint in ventilating their anger' and grief. However, some spontaneous anti-Indian demonstrations were staged in Dhaka and elsewhere in Bangladesh. No Hindu was killed in these demonstrations. Bangladesh Government took unprecedented precautionary measures to check any untoward happening. Nevertheless RAW stooges based in the United States shamelessly launched vigorous anti-Bangladesh campaign about alleged excesses against Hindus in Bangladesh. Scores of anti-Bangladesh demonstrations took palce in front of the UN building and the Bangladesh Consulate in New York. Anti-Bangladesh slogans condemning so called communal and racial violence were raised in these demonstrations.

Memorandums were submitted to the UN Secretary General, Bmlgladesh Consulate General and other Human Rights organisations levelling charges against Bangladesh Government on its supposed failure to protect the rights and properties of the Hindus. RAW is also using local Indian magazines including News India and India Abroad etc. for anti-Bangladesh propagation. Distorted facts and figures are published in these magazines for furtherance of RAW's objectives.

Special articles are written by RAW hirelings to defame Bangladesh. RAW stooges project Bangladesh as a worthless country whose people are. fanatics, uncivilised and devoid of civic sense. Thus what they say about Farakka or Talpatti or about Indian involvement in CHT affairs is senseless.The above are some examples of RAW's anti-Bangladesh propaganda campaign abroad. It is imperative for Bangladesh Government to take appropriate steps to thwart RAW's propaganda.

ERECTION OF BARBED WIRE FENCE ALONG BORDERS

India is pursuing an ambitious and higly expensive programme for erection of barbed wire fence along the Indian border with Bangladesh. It is amazing that India has embarked upon the programme at a time when the trends all over the world are to remove physical barriers between nations and the people. The removal of infamous Berlin Wall is an example of same phenomenon. The walls, fences and other physical barriers are constructed to protect a country against external threat. Question arises as to what threat Bangladesh poses to India's security? Obviously there is no threat to India from Bangladesh. Then why India is implementing the programme despite very high costs?

The usual explanation given by Indian authorities is that they want to seal India's borders with Bangladesh in order to check illegal infiltration. However, their real aim is indeed sinister. The Indians want to convert the borders into a one way valve whose control will be with them. This will enable them to pursue their various nefarious designs against Bangladesh with ease including 'push back', launching of terrorists, saboteurs, spies, organising one way smuggling and drug trafficking etc.

The erection of wire fence is being undertaken in a phased but rapid programme. Work is already in progress in various sectors of Indo-Bangladesh border. Reporting about progress of the work in Dinajpur, Thakargaon and panchagarh areas, Daily Janakantha wrote: "Day and night work is going on to build metalled roads along the border. Similarly erection of wire fence is also being undertaken at rapid speed. Already 507 kilometre long fence has been erected. Moreover Indian Border Security Force (BSF) has instructed the farmers who have lands along the border not to cultivate crops which grow tall enough to obstruct observation. The expenditure on construction of the roads and erection of fence is estimated to be Rs 468 crores(Janakantha : May 1, 1994).The Daily Sangram reported in its issue of July 27, 1995, "Fence is being erected in three rows from Sona Moque topposite Chakpara) to Karimganj-Kaliganj area. Besides, observation towers and metalled road have also been constructed withhin 200 yards of the border."

The Daily Al Mujadded reported, "The barbed wire fence in Jibannagar sector of Chuadanga District has three rows. The first row is 9 feet high fenc,e mounted on angle iron pickets. The middle row is 4 feet high straight fence and the last row is 6 feet high coil fence. The centre row is electrified. The rows are 3 feet apart from each other. A 20 feet wide metalled road has been constructed at a distance of 10 feet from the fence. Steel gates have been fixed in the fence at each 2 kilometre distance"(Al mujadded : May 14, 1995)."Along Indo-Bangladesh border in Satkhira sector, India has constructed barbed wire fence all along the border. The fence has three rows. The middle row shall be later electrified. Observation towers are being built at close intervals(Mujadded : April 10, 1995)."India has also erected barbed wire fence at Garo Hills area, Sylhet-Maulvi Bazar border and along border of Tripura state(Daily Janata : May 12, 1995).

Erection of fence along the border is part of a bigger consipiracy by RAw to cripple Bangladesh. There are repotts that after construction of barbed wire fence, push back operations will be undertaken with maximum vigour as RAW believes that after erection of fence, it will be impossible for p.eople pushed into Bangladesh to return. Similarly there will be tremendous increase in infiltration of Indian agents and smuggling of all types 'of goods including drugs into Bangladesh.

Illegal Occupation of Bangladeshi Land

India has illegally occupied Bangladeshi land in many border areas. No where in the world a country occupied a neighbouring country's land the way India has Bangladesh's land. India is determined to perpetuate her illegal possession of such land. That is why in the past 23 years of Pakistan and in another 24 years of Bangladesh, India never seriously sat down to solve the border issues. At many places she did not even give areas allotted to Bangladesh by Radcliffe Award.

India has been persistently refusing to undertake joint survey and demarcation of borders. In many border areas Indians evenremoved old border pillars and encroached upon Bangladeshi territory. In areas where middle stream of the river is recognised as the border, India diverted the river course towards Bangladesh territory and grabbed Bangladeshi area. The illegal occupation of Bangladesh's land by India has taken place at over one hundred places. However, due to paucity of space, here I whould' give details of only a few cases to inform my readers about Indian manipulation and tactics in this regard.

"India is in adverse occupation of 200 acres of Bangladeshi land at 'Berubari sector in the district of Jalpaiguri. A survey team consisting of members from both countries visited the area in 1991. Sensing that solution to the problem may be found by the team, RAW decided to put hurdles in the way of the team. It manipulated sentiments of the people in Indian area by spreading rum our that if the team is allowed to work the area would go to Bangladesh creating problems for the people of the locality. Some political parties of India including BJP and Forward Block were instigated to launch movement against survey work at Berubari area and the possible handing over of some land to Bangladesh. That forced the survey team to abandon its work and leave the locality without accomplishing its task(Daily Bhorer Kagaz : November 11, 1994).

The Daily Sangram reported, "Due to India's obstinacy and hegemonism two border points in the areas of Rajshahi District have remained unsettled for a very long period. These two areas are Chandansahar of Charghat Police Station and the other is the Nirmalchar of Godagari Police Station. Although these areas are rightly in Bangladesh, yet the Indian BSF has always been putting pressure to forcibly occupy them. In October 1992, the Indian BSF supported some India citizen to attack Bangladeshi farmers working in Nirmalchar area. At one stage BSF forcibly occupied the area and dug some bunkers as well. However, on being approached by BDR they left the area. But the Indians have not yet agreed to conduct a joint surveyw settle the issue for good. A source maintains that the Indians intend to forcibly occupy the area(Sangram : March 25, 1994).

"The Indians claim that since Padma River is the border line, therefor, Chandansahar shore belongs to them. However, they forget that in this region the river course keeps on changing. It does not mean that border line will also change. Decision about the ownership of shore should be made based on old records of river course in 1947. India, however, is not ready to accpet it. Bangladesh Government sources stated that the Indians have been requested time and again to settle the matter but they did not respond at all. As a result the Bangladeshis living in the area are under constant threat of intimidation and fear from the Indians(Daily Sangram : March 20, 1994).

"In the Thakargaon border area, the Indians have been trying to forcibly take away 90 acres of land from Bangladesh. In January, 1994, some Indians entered into the area and started preparing a tea garden there.Although BDR and BSF jointly settled the matter, the Indian farmers once again on 2 February, 1994 entered the area and started work for the same end. This time the Indians supported by BSF, managed to make a tea garden over an area of one acre of land(Daily Bangla Bazar Patrika : February 20, 1994).

The Daily Ittefaq reported, "Of the 180 miles border with India in the District of Panchagarh, 18 miles are demarcated by river. The river course has been closing in towards Bangladesh every year due to floods and picking up of stones by labourers by breaking the river banks.This has resulted in creation of new shore. India has occupied newly formed shore though these do not fall into her territory. When contacted, the relevant department of Bangladesh Government informed that the Indians do not turn up on the appointed day and time for deciding such border demarcation issue. They keep on dragging these cases. Knowledgeable sources maintain that motivated by their evil intention to keep Bangladesh's land under illegal occupation, they do not send their survey team to settle these matters(Daily Ittefaq : Sepember 19, 1994).

"On December 19, 1993, the Indian BSF personnel forcibly occupied 20 square kilometre of Bangladesh's land adjacent to the Sunderbans of the Kaikhali and Ramjan Nagar Unions located at the end of Satkhira District. They sank down more than 100 boats of Bangladesh, kidnapped some fishermen and threatened the others not to re-enter into the 10cality(Inqilab, December 27, 1993). "India has kept under forcible illegal occupation 1500 acres 'of land at the Taindong and Asalong villages of Matiranga Police Station in the Khagrachari Hill District.

In 1986-87, people of these villages, after being attacked by Shanti Bahini guerillas, left the area due to lack of security. The BSF occuppied the said area comprising of 1500 acres. BSF has set up two camps in the said land. They have also constructed metalled road and erected electric lines. When in May 1995, the Indian farnlers, on being prompted by BSF, started cultivating the said land the Bangladeshi farmers objected. However, BSF still continues to occupy the said land(Daily Janakantha, May 25 1995).The Daily AI Mujadded reported, "India has forcibly occupied 2300 acres of land along he greater Kushtia border area.

It is located between pillar No. 126-127, pillar No. 132-133 at the Kathuli border, in the Isakhali border from pillar No. 122-15 to 122-65, and of Kushtia District at pillar No. 153-25 to 154-45. Besides, 300 Bigha of Bangladeshi land is under forcible occupation of India, between main pillar No 80 and 82 at the Baradi border of Damorhuda Police Station and between pillar 89 and 91 in Jhakurpur border of District Chuadanga(Mujadded : June 14, 1995)."

A Media Syndicate report published in The Daily New Nation stated, "India has occupied 2000 square miles of Bangladesh's territory by constructing dams, barrages and other structures on rivers entering Bangladesh.

India has constructed 25 dams and 40 structures on 25 rivers. These structures have caused loss of water in rivers. By putting up pillars and fences in dry areas India is claiming sovereignty, over 2,000 square miles of land that belongs to Bangladesh(New Nation: October 1, 1995).

South Talpatti Island

A glaring example of India's forcible occupation of Bangladeshi land is the case of South Talpatti Island. The Island is located at a distance of 6000 metre from the mainland shore of Bangladesh as measured by aerial survey obtained from remotely sensed imageries.

The Chandbaria range of Bangladesh's Forest Department in the District of Satkhira is the nearest landmass of Bangladesh. The border in this region is based on River Hariabhanga whose main course lies in the west of the Island. Thus the Island forms an integral part of Bangladesh. The Radcliffe Award had fixed mid river course as the border between India and the then East Pakistan. The same formula was inherited by Bangladesh. Based on the said principle, the ownership of South Talpatti Island should lie with Bangladesh. In a survey map made in 1967 by Survey of Pakistan, the Island was shown in East Pakistan (now Bangladesh). The Survey of Bangladesh map issued in'1972 had also shown the Island as part of Bangladesh. But India is not prepared to accept the said reality.

On May 9, 1981, India landed her armed forces on South Talpatti Island with the support of three warships of Indian Navy. Since then India is maintaining her illegal occupation of the Island. Bangladesh has sent many proposals to India to settle the ownership issue of the Island. Bangladesh also proposed a joint or international survey. But India does not agree to hold any survey work. They know that such work will establish that the Island belongs to Bangladesh. Disregarding all principles, India has resorted to the jungle rule of 'might is right'.

Establishment of Bangladesh's right over the South Talpatti Island will add 25,000 square kilometers area to Bangladesh. Besides, the naval and strategic importance of this Island to Bangladesh is enormous.

The shallow sea shore and beach around this Island can be developed for tourism purposes. Moreover geological surveys have indicated deposits of huge minerals like iron ore, Manganese, Aluminium etc. in the off shore of the Island.

Motivated by her desire to grab rich resources of the Island and also in pursuance of her evil design to re-establish Akhand Bharat, India has kept her illegal hold on the Island. It is well known that Bangladeshi fishermen were the first to reach the Island.

They had established fishing camp on the Island in the past. In 1974 Sheikh Mujibur Rehman gave permission to six foreign companies for oil exploration in the area. These companies carried out exploratory work in the off shore of Bay of Bengal. Indians raised their claim only after 1974. Question arises that if the Island was India's territory, why did not she claim it before 1974. It may also be noted that India established her possession of the Island only in 1981.

India is gradually consolidating its hold over the Island. In this regard the Daily Meillat reported, "India has strengthened her armed position on the Island. India is already negotiating with foreign companies to grant them right for oil and gas exploration around the South Talpatti Island.

The concerned Indian Ministry has already invited tenders for the work."The issue of unsettled border lines between Bangladesh and India is indeed a major reason for constant trouble and souring of relations between the two countries. In 1974, Sheikh Mujibur Rahman and Mrs Indira Gandhi had signed a treaty for solving the unsettled border issues. The treaty signed on the May 16, 1974, was published in the Bangladesh Gazette on the November 28, of the same year.

The border treaty was later incorporated into the 3rd amendment of our constitution. In this treaty it is stated that all survey works would be finished by December 31, 1974 by both countries. It was also provided in the treaty that the map (new) would be prepared by May 21, 1975 and a jointly approved map will be published by December 31, 1975.

Details of disputed border lines were also mentioned in the treaty signed by both parties. After signing the treaty, the survey work began in some areas. But after Sheikh Mujib was killed on August 15, 1975, the Indian Government started to show unwillingnes to make any further progress in the matter on flimsy grounds"(Mujadded : June 14, 1995).

According to international law India is bound to obey the terms of the treaty, notwithstanding change of the' Government in Bangladesh. But India is refusing to honour the treaty. She has also been flouting the terms of the treaty and is constantly endeavouring to grab Bangladesh's land illegally and unlawfully through the use of sheer force.

Late President Ziaur Rahman had adopted a bold stance on the issue of Talpatti Island. He also pursued diplomatic offensive in the matter. But the subsequent Governments of Bangladesh failed to do anything like that. Analysts believe that the RAW's hold on the subsequent Governments of Bangladesh has increased with the result that national interests and sovereignty no more take first preference. That is the reason that Talpatti Island and many other lands in border areas remain under illegal Indian occupation.

Despite Bangladesh's gesture of giving two square miles of area of Berubari in exchange for Tin Bigha, Indians have failed so far to show any accommodation on the issue. Observers believe that RAW is a major player behind the scene which has been persuading Indian Government not to make any settlement of border lines and Talpatti Island issue etc. RAW is waging a war of attrition against Bangladesh and wants to use every possible mean to cause suffering and loss to Bangladesh. RAW's aim is to demoralize and pressurize Bangladesh through every conceivable way so that realization of their end game is facilitated.

11

RAW in Secret War Against Pakistan

RAW had been created by India in 1968 with the sole aim of destabilizing East Pakistan, subverting its leadership and taking it in its total control and creating suitable conditions for the intervention of Indian armed forces. After Awami League won landslide victory in 1970 elections, Mujibur Rahman became a hostage in the hands of extremist Bengali leaders like Tajuddin.

He became headstrong and arrogantly defied the writ of the federal government and made a parallel government in Dacca. He stoked Bengali nationalism and allowed his thugs to undertake mass slaughter and rapes of non-Bengalis so as to provoke Gen Yahya Khan to launch a military operation.

When the carnage was taking place against hapless non-Bengalis, Biharis and pro-government Bengalis for 23 days starting 2 March 1971, the Army had been confined to barracks. A blackout was imposed upon state media on the silly pretext that it would spark a counter action in West Pakistan.

Once Operation Searchlight was ordered on 25 March to save the lives of stranded civilians and to save the federation from splitting, it gave an excuse to the Bengali leaders taking refuge in India to start a civil war fully aided by Indian military. Massive propaganda war was unleashed by India supported by western media to demonize Pak security forces. They were painted as blood sucking monsters and rapists, while the rebels were eulogized.

After forcibly severing eastern wing of Pakistan and making it an independent country in December 1971, RAW started meddling into internal affairs of truncated Pakistan and worked on provinces which were economically less developed and had grievances. It first worked in real earnest in rural Sindh where the poor segment was being exploited by Waderas and where there was sizeable presence of affluent Hindus who had influence over Sindhis. GM Sayyed provided RAW a toehold and other Sindhi nationalist parties aspiring for Sindhudesh were taken on board. Rural Sindh posed first major challenge to the federation under Gen Ziaul Haq in 1983 when Movement for Restoration of Democracy was spearheaded by PPP under Mustafa Jatoi. The movement was supported by India's Lok Sabha and Indra Gandhi hosted All Sindhis conference in New Delhi.

When the movement couldn't make any headway since it remained localized to interior Sindh only, RAW then shifted its emphasis to urban Sindh with focus on Karachi with the help of newly established MQM. Grievances of Urdu speaking community against Sindhis and latter against former were fomented by RAW. Antagonism between old and new Sindhis was stirred up to create rural-urban divide. Among the MQM, some toyed with the idea of Jinnahpur and started working on Mujib's model to control Karachi through terrorism for which it created a militant wing.

In the Frontier Province, Indian leaders maintained their contacts with the help of NAP that was aligned with Congress and had espoused Pakhtunistan. The party that was banned by Bhutto in 1973 was renamed as ANP in 1978, but it maintained its contacts with India. The nationalist political leaders of Sindh and Frontier gave a commitment to India that come what may they would not allow Kala Bagh dam to be constructed.

RAW made its inroads in Balochistan during 1973 Baloch insurgency via Afghanistan. In concert with KGB and KHAD, it started subverting the people in Balochistan and the trio actively supported Baloch insurgency which raged till 1978. India got aligned with Soviet Union and Afghan regime in 1980 and the intelligence agencies of the three countries together with Al-Zulfiqar, a militant wing of PPP, carried out intensive acts of sabotage and subversion against Pakistan till the defeat of USSR in 1989. In late 1980s, RAW established link with a group in southern Punjab, which was instigated to seek establishment of Saraikistan.

India's covert activities received a big boost after 9/11 since the US, Israel and western world ganged up against Muslim world and India was promised all out support to defang nuclear Pakistan. It became easier for RAW agents to increase its presence in various parts of Pakistan since the latter having become an ally of USA had opened its doors for its intelligence agencies. CIA after consolidating its position in targeted regions facilitated entry of RAW agents as well. Indo-Pak peace treaty in 2004 opened further vistas for India to befriend desired audiences and to promote Indian culture particularly in Lahore, the heartland of Punjab. Inroads in print and electronic media enabled the US, India and Israel to build perceptions of choice and to defame Pak institutions. Balwaristan movement seeking independence of Gilgit is RAW creation and so is TTP in FATA. Phenomenon of target killings in Karachi from 2008 onwards, which provoked the PPP and ANP to establish militant wings to counter MQM's vandalism to control port city, had the backing of RAW. 3500 people died in senseless target killings.

Having spread its tentacles all over Pakistan, India is now in an advantageous position to heat up any of the trouble spots suiting its design of covert war and that too without any fear of retaliation from Pakistan inside India because of Pakistan's policy of appeasement. Pakistan has been put on the defensive because of sustained propaganda war by Indo-US-Western nexus. In the aftermath of Mumbai carnage, both India and USA threatened that Pakistan will be held accountable in case of any future act of terror taking place in India. Pakistan has become so cautious that it is not even supporting Kashmiri freedom fighters seeking right of self determination.

RAW is deeply involved in Balochistan, FATA, southern Punjab and Karachi. This time it is helped by USA, Britain, Israel, Germany and Afghanistan. Its nine years sustained efforts have resulted in transforming Baloch insurgency into a separatist movement. The rebellious Baloch Sardars and nationalist leaders initially voiced political and economic grievances and demanded provincial autonomy and greater share in mineral resources of the province. Once most of their demands were met and mega projects launched to address the inequities, they started demanding independence of Balochistan. Those demanding separation are not more than 3%, while the rest of Baloch people are highly patriotic.

In FATA, TTP is aligned with Afghan Taliban and has an agenda of imposing Shariah in FATA. Since some militant groups in North and South Waziristan are pro-Pakistan and anti-US, and that none seek independence from Pakistan, hence the US and western world keep jabbing Pakistan to kill anti-US Pashtuns. At the same time, TTP is pricked to continue with its militancy and prevent normalization of FATA and Khyber Pakhtunkhwa where stunt of Pakhtunistan has not died down.

The stance of USA in Balochistan is quite the opposite. Here even slightest retaliatory action by the paramilitary forces is drummed up as violation of human rights. It is falsely claimed that the Army has undertaken a military operation and that intelligence agencies are involved in kidnappings of Baloch dissidents. Baloch rebellious groups like BLA, BRA and BLF trained and equipped in Afghanistan are involved in sabotage activities and in target killings of non-locals, Hazaras and pro-government Baloch. They are terrorizing peaceful citizens and poisoning the minds of the Baloch youths against non-locals, central government, Punjab, Army and ISI. They are told that despite their province being the largest and rich in resources, it is most under developed. Blueprint is the same as was practiced in erstwhile East Pakistan. Tabling of a bill by three US Congress members is an attempt to internationalize Balochistan issue.

Notwithstanding grant of MFN status to India and to make India part of Afghan Trade Transit to boost mutual trade, it must not be forgotten that India has not stopped its covert war against Pakistan to destabilize and denuclearize Pakistan and to detach Balochistan. India has yet not closed its training camps and Pakistan specific consulates in Afghanistan, nor stopped its support to Baloch dissidents. RAW is a notorious rogue outfit, which is chiefly responsible for disturbing the peace of all neighbours including Afghanistan. Indian leadership instead of bridling its insidious activities impishly blames ISI, which is not exploiting insurgencies in 19 states of India.

Indian leaders are trying to become overly friendly since its chief benefactor USA itself has got entangled in host of problems and is set to quit Afghanistan. We know how cleverly India had deceived Pakistan by signing peace treaty in 2004 and promised to solve all core issues through composite dialogue, but used the period of friendship to promote RAW's secret war to undermine Pakistan and to build dams over three rivers to choke Pakistan. Last seven years have been nightmarish for Pakistan.

Should we again be misled into believing that this time India is genuinely interested in peace and friendship? If so, what is the basis of optimism? What practical steps India have taken to allay our legitimate fears, or to resolve outstanding issue of Kashmir, which bedevil Indo-Pak relations? Or our leaders have simply got mesmerized by preachers of RAW sponsored Aman ki Asha? For 64 years Pakistan has been ceding ground to win the friendship of India but couldn't change the mindset of Brahman rulers. If India is genuinely interested in friendship with Pakistan, it must prove it by actions and not by deceptive and hypocritical sweet talk.

PAKISTAN: RAW FACTS

RAW used the growing disaffection of the Bengalis in East Pakistan to build up the foundations of Bengali separatism. It lay the ground for the Indian military entry into what was then East Pakistan by building up, training and arming the Mukti Bahini. RAW's failure lay in being unable to lead the Mukti Bahini and the Bengalis to the 'natural' conclusion of their struggle against Pakistan - as planned by India - that was a union with India. Instead, perhaps RAW overkill helped the Bangladeshis into seeking a gradual distancing from India so that in the end the creation of Bangladesh only helped to reassert the relevance of the Two-Nation Theory. This has not prevented RAW from focussing its activities in Pakistan - keeping a careful eye on all developments within Pakistan's domestic polity. In Pakistan RAW has had a multi-pronged strategy using the Indian media, abetting political subversion and actively developing a terrorist network which becomes operationalised within Pakistan as and when RAW feels the time appropriate. That is why there has been a gradual transformation of simple political dissent into a violent form of political polarisation and subversion.

That the opportunities have been provided by the local political machinations cannot be denied - but RAW has been quick to take advantage and introduce an ever-spiralling element of violence within the political discourse and conflict that prevails in all the South Asian countries.

So the fringe minority elements within the Pakistani polity have found themselves receiving RAW largesse which has allowed them to build up their militant resources. RAW has insidiously played the tune of a 'common subcontinental heritage' despite the fact that, barring the period of British colonialism there never was a 'united India' except in the minds and dreams of Hindu chauvinists. RAW has of course varied its tactics keeping in mind the groups it was seeking to bolster or influence.

Before the Soviet intervention in Afghanistan altered the dynamics of the whole Pakistan-Afghanistan relationship, RAW sought to nurture the Pakhtunistan issue clandestinely even as the Indian government sought to overtly cosy up to certain members of the political elite in the NWFP.

When RAW saw an opportunity in Balochistan in the form of Baloch discontentment - especially in the face of ZA Bhutto's dismissal of the elected government of that province - it moved in fast to play up the issue of Punjabi dominance. While the Baloch insurgency may have had indigenous roots, it was nurtured and sustained with external aid and assistance and RAW's trademark was clearly evident through the Afghan route.

While the Pakhtunistan issue gradually faded away with the altered realities on the ground and the Baloch insurgency was isolated and put down, a more serious crisis was brewing for Pakistan when General Zia imposed Martial Law and then hanged ZA Bhutto, who came from Sindh. The Sindhi distress at this act was seen by RAW as an opportunity to exploit. Already there had been disaffection in rural Sindh when One Unit had been established in 1956. Rural Sindhis had also become increasingly wary of what they saw as a high concentration of Urdu-speaking Muslim migrants from India in Sindh. The language issue further aggravated this division within the population of Sindh while self-serving local politicians constantly played the theme of underdevelopment and deprivation of Sindhi and the Sindhis. With Bhutto's removal from power by the military and his subsequent hanging, all the Sindhi discontent gelled together. So for RAW Sindh became an increasingly fertile ground for their seeds of violence and terrorism.

RAW also upped the violence ante by brazen acts of terrorism which had the RAW hallmark all over them. So blatant was RAW in its earlier terrorist activities in Karachi that the Pakistan government could not turn a blind eye to these activities - especially since they also involved the Indian Consulate in that city. When the links between the Consulate personnel and ethnic dissidents became too obvious to ignore, the government had to have the Consulate closed. However, this closure has not stopped RAW from infiltrating ethnic groups so that the ethnic conflict in Sindh continues to remain highly militarised and there are constant links emerging between RAW and the acts of terror conducted by these groups.

Nor has RAW ignored Punjab in Pakistan. Again, exploiting opportunities provided by indigenous developments, RAW has ensured that socio-political conflicts get more violent and thereby more polarised. By turning political conflicts into military conflicts, RAW hopes to make them more intractable. In Punjab, RAW has made inroads into the sectarian groups so that there are now heavily armed and trained terrorists who have infiltrated these groups. Random acts of terror are also conducted within range of Punjab's urban centres so that an element of fear creeps into the polity at large. The number of bomb blast incidents in Pakistan for the year 1998 (given below) show how violent the Pakistani polity has become - and RAW can claim credit for a large part of this development.

Moving beyond physical terrorism, RAW has also inundated Pakistan with inflammatory literature to play on the sentiments of minority groups, as well as sectarianism and ethnicity. While Pakistan has been evolving its democratic ethos after the last bout of military dictatorship which spanned over a decade, RAW has attempted to take advantage of the multiple levels of political dissent that any democratic polity - including India - has to contend with within the framework of democracy.

RAW has also sought to undermine Pakistan's external image, especially in the West - taking advantage of the Western phobia of 'Islamic fundamentalism-terrorism'. The Indian media and responsible leaders have orchestrated the campaign to hold the ISI responsible for India's troubles in her northeastern states and East Punjab. In April 1995, the Indian Army Chief of Staff, General Chaudhri, went public in his accusation against the ISI whom he held responsible for the guerrilla movement in the northeast as well as the troubles in Indian-held Kashmir.

RAW also conducted a well-organised campaign to try and have Pakistan put on the terrorist watch list and, for a while, in the early 90s there was a fear that Pakistan would be declared a 'terrorist state' by the US.

At present, RAW has launched a new offensive against Pakistan to try and counter Pakistan's successful efforts to expose Indian human rights violations in Indian-held Kashmir internationally. With the Indians desperately seeking a way out of their Kashmir quagmire, RAW has started a new propaganda offensive focussing on what it refers to as Pakistan's 'Proxy War' in Kashmir. With the rise of the Madrassah culture in Pakistan and its links to the Taliban in Afghanistan, RAW is trying to create a linkage of this with the Kashmiri freedom fighters in Kashmir. One does not need too much intelligence to see where the RAW campaign is leading. But it should at least make the Pakistanis realise that India is seeking a way out of its untenable position in Kashmir.

All in all, when the international community is increasingly condemning overt war as an instrument of state policy, India has already sought an alternative, indirect and covert mode of warfare through RAW which seeks to destabilise and weaken the states of South Asia from within. Given the changing nature of war one should be prepared for more RAW activities since politics is increasingly becoming the 'continuation of war by other means' - and RAW has evolved the expertise on 'other means' in South Asia.

COVERT OPERATIONS AGAINST PAKISTAN

India has limited experience of covert operations against its western neighbour. Clandestine electronic intelligence gathering and aerial reconnaissance gives it some part of the picture. Additionally, it has gained experience in exfiltrating agents into POK into the training camps and this has helped in keeping tabs on activity in the POK region. But India's ability to influence events in Pakistan itself is limited. Small-scale moral and material support does go out to ethnic groups in Pakistan and some effort is made to infiltrate 'spies' by Military Intelligence in the border regions. But the larger picture is still dependent on Technical/ electronic intelligence and even here capability to tap landlines in Pakistan is limited. Human intelligence from within Pakistan is much less than required. It is however suggested by one source on the Internet that RAW has over 35,000 agents in Pakistan.

Therefore, plans have to be made to covertly gather intelligence from inside Pakistan to help in foreign policy objective of breaking the monopoly of the ISI and army over Pakistan. India has to have a plan of action to destabilise Pakistan, its economy and society, to the extent that it gives us leverage in foreign policy terms. It must however, be clear that it is not in India's interests to have a disintegrated Pakistan. The aim is to break the stranglehold of the intelligence agencies, the bureaucracy and the military in Pakistan.

The proposed strategy is to undertake covert passive and active measures against Pakistan including, *psy. ops,* disinformation, strikes in rear areas, border raids and so on. The last may include strikes by special forces against training camps in POK and Northern Areas. Such operations require a clear national will and motivation. They also require sustained funding from the political leadership and it requires highly trained and motivated manpower to execute this task. *The objectives of such operations are to first, penetrate Pakistani society and its institutions of power. Obtaining intelligence from within the establishments of power is the main aim. Associated with this is the second, long-term aim of breaking the stranglehold of the main power brokers in Pakistan.*

For the above mission, there is a need to create, to start with two sets of teams. First, a plans section and second, an operations section. The first will draw up the plans and stages for operations against Pakistan. Also recruiting agents for the task and training them should be completed in about six months. Planning from conception to actual operations should take 8-12 months, depending on the resources and manpower available. Once this is achieved the operations begin.

The operation can be politically cleared by RAW and then a separate section can be raised under the Director RAW, drawing the best talent from within and other agencies including the army, all volunteers of course. Known as Plans Directorate (PD), this will mastermind the operation. Personnel for Strike Directorate (SD) should be picked from SSB, Army para-commandos and NSG. For the agents and sleepers recruitment has to be from outside. The need to know principle has to be strictly followed. Accountability should be from Prime Minister, National Security Advisor, to Director RAW and to PD head.

The entire operation can be divided into three phases. First phase is penetration and setting up of networks. Second phase is to begin operations in rear areas. Strikes against soft targets in Sind, Baluchistan and Northern Areas. These will make the enemy react and make them sensitive.

The main aim is to infiltrate and subvert Pak institutions, the police, communication network and other important organisations. To this end, it makes sense to penetrate Pakistan from two sides, from outside the sub-continent and from POK. Then, in the next stage, focus moves to organising and training subversive elements. These groups would target communication and logistic lines inside POK. Threatening the Mangla dam or Kahuta for instance, would be a major psychological factor. Third phase is hitting hard targets. First, to carry out disinformation campaign carried out by agent provocateurs and others. A few expendable agents are used to plant false information on police and intelligence agencies. At this stage, the strike teams begin hitting hard targets like economic centres, financial markets (like Karachi), ports and the like. Military installations like ammunition dumps, communication centres, airfields can be targeted in case a war breaks out. Otherwise the aim is to penetrate and gather intelligence with the intention of breaking the stranglehold of the military over Pakistan society.

The mission entails sending two teams of men (and women) into Pakistan for two separate missions. This first team will consist of sleepers, and agent provocateurs, and the second team will be the strike force. The sleepers and agents will consist of network builders and specialists who will recruit local people for both disinformation and strikes in the rear areas of Pakistan. The agent provocateurs will be placed such that they can engage in disinformation and activate sources within who can be expendable when required. The second team is for strikes against key targets of economic and military importance.

Both teams will have to be infiltrated on the basis of genuine documents and placements in Pakistan. Identities can range from middle level social workers, journalists and political workers. Crucially, the persons selected for this job will have to be extensively trained for operations in living off the land in Pakistan. One may at this stage refer to the operation conducted by the Israeli Mossad in Syria aimed at the KHAD party. The agent was an Israeli but was trained to behave and live like a Muslim. Therefore it should not be difficult to train an Indian, Hindu or Muslim to live in Pakistan. The key lies in creating excellent and plausible identities. Training has to be rigorous and thorough. Training will also be common initially and then separate and compartmentalised for the two teams - one, sleepers and agent provocateurs and two, demolition teams that will serve as strike teams in the rear.

While both teams need training in explosives, weapons and communications, each has a specific role, which calls for intensive training. For the sleeper the aim is to penetrate Pakistani society and build a local network for further infiltration. The important thing is therefore for these members to be 'Islamised.' For the strike teams it is going further afield, into Sind, Baluchistan and Northern Areas. This means differing identities and covers; identities are important. The first team has to consist of "Pakistani nationals." Punjabi or Sindhis, who are from abroad, can be targeted. Some thought can be given to recruiting people from the Asian community in UK. Some members can be selected from Uzbekistan or Turkmenistan. For those coming in from the west, penetration point is obviously the United Kingdom, USA or Europe. Even the Middle East can be used. In other words, the potential areas for recruitment are large and this needs to be analysed carefully.

The strike teams could penetrate from Afghanistan or Tajikistan. The NWFP is full of Afghan refugees and the Northern Areas have Turks, Tajiks and Uzbeks. Finding these persons who are reliable and capable is going to be the tough part, but not impossible.

Another option is to exfiltrate these teams from Indian soil. Therefore, the strike teams could penetrate from northern Kashmir. The northern areas of Jammu & Kashmir bordering Gilgit and Baltistan, mainly Kargil is a Shia dominated region, this could also serve as a recruiting base.

There are also traditional routes that could be used for exfiltration into POK. Additionally, India can find local recruits from amongst the Turks and Tajiks in these regions who could act as couriers and informants. Similarly, in POK the Gujjars and Bakkarwals can be recruited for the same purpose. Seasonal migrants and sheepherders come from the Northern Areas, Chitral and Koistan areas. They can also be recruited. The aim of penetrating the strike teams is to get them to strike in the rear areas of Pakistan. The entire infiltration process of both teams, will take at least one year. Then it is a slow task of making inroads into Pak society.

So it is clear that mission objectives of both teams are different and require different infiltration and exfiltration routes. As suggested above, the first team cannot jump off from India; the second might, given the right circumstances. The sleepers and agent provocateurs are always in danger of exposure. Thus their exfiltration will depend on their own resources and on the overall command and control of the mission. If mission control did want to abandon the task, then alternative exfilitration routes will require to be planned. The strike teams consisting of individuals, usually two to four must have routes of escape, and this is possible in those areas bordering India. It is thought best, diplomatically, at some stage to use Tajikistan as the exfilitration route for those employed in the Northern Areas and surrounding regions. Those in Baluchistan, Sind and NWFP will have to use Afghanistan or get to the Arabian Sea for exfiltration.

The first team should consist of 10-20 persons, both men and women who will be sent to Islamabad, Karachi, Rawalpindi and Lahore. The second team, consisting of only men will go to towns in Sind, Baluchistan and NWFP, as also the areas of Gilgit and Chitral. Numbers will depend on how much rear area attack is envisaged, perhaps initial numbers may be 8-10 in groups of two, to each region. The entire operation has to be planned keeping in mind that the initial period of infiltration and consolidation should be about four-six months. And then once the strike teams have got into place, situations can be created.

The aim of the first team is to settle in Pakistan, infiltrate society, make inroads and build networks for information gathering and find niches in high places. The agent provocateurs will be expendable and need to be locally recruited, though for the sake of authenticity, one or two sleepers need to be expendable also. Key areas of penetration are the bureaucracy, the army, ISI, elite middle class and media. After a lull of a year, information should be leaked, through available means of the existence of these agent provocateurs, who will have in their possession, propaganda material, disinformation material and generally stories that lull Pakistani intelligence into a sense of complacency that they have achieved a coup. Further, as the Russians so successfully did with the CIA and MI-6, the agent provocateurs should talk of moles and other sleepers in the Pak military and intelligence establishment to create internal dissension.

Once this leads to internal turmoil, it is time to activate the strike teams who will launch rear area attacks on installations of economic and military value. This will lead to further conspiracy theories being floated, leading to political uncertainty. The risk that this entails is that it will weaken the democratic process and possibly lead to military rule in Pakistan, but that is a contingency that India has to prepare for.

The aim of strikes is to create a situation where the establishment is caught in an over-reach situation. The rear areas, which are under the administrative control of Pakistan, but given its tribal and ethnic composition are quite independent, can be suitably engaged by India for its own purposes. Strikes will mean deploying more police and para-military troops to these regions and will stretch the Pak establishment.

Surprise and timing are an important element in these missions and those in the strike formations should be constantly on the move, working from small towns, crowded market places and religious places. Rear area operations are to succeed by strikes in a few important cities and installations of economic value such as oil fields and tankers. Karachi city and its port is an example. Period of operations will depend on the nature of Pakistani response; if they do according to plan then teams can be exfiltrated. If they need to be in for some more time, then further deception plans need to be put into operation.

The duration of operations in most cases is stretchable. The first phase may take anything between one and three years, while the second phase will be carried out in synchronised fashion for a month or two in one place and then move on to another target. After a respite of a few months it will be time to carry out operations again. The long-term objective is to build information networks in Pakistan that will provide a human intelligence source for India and additionally it will help us make inroads into the military establishment and bureaucracy.

Along with the strike missions carried out by the second team, plans may be made to carry out heliborne operations and SF strikes in the border regions against militant training camps and other suitable targets. Forces for this are already available in the form of the National Security Guards and Indian Army para-commandos. These can be configured and trained for operations in a near-war like situation or simultaneously with the covert strike missions outlined above. All this of course requires a national security policy that seeks to engage and contain Pakistan in foreign policy terms.

To aid this process and to improve intelligence gathering on Pakistan the above covert action plan is commended to the Indian decision-maker and political leadership. The Indian diplomatic and security response to Pakistan sponsored terrorism has often been to fight fires within after the incidents have taken place. Additionally, there is a dialogue track, which seeks to engage Pakistan in a meaningful bilateral exchange. But little thought is given to the possibility of engaging in covert operations on a scale suggested above and its usefulness as a tool of diplomacy. The intention as stated above is not to help the disintegration of Pakistan but to engage Pakistan in a battle in which the key players, the army, ISI and bureaucracy will get embroiled to an extent that it creates conditions for, first Pakistan's disengagement from the sub-continent and second, may help the process of bilateral dialogue.

RAW: INDIA'S INTELLIGENCE SERVICE AND ITS INVOLVEMENT IN PAKISTAN'S UNREST

India's intelligence agency, the Research and Analysis Wing (RAW) is extensively involved in Pakistan's, and neighbouring countries', affairs along with other foreign spy agencies such as the Israeli Mossad and the Afghani NDS. Based on the principle of conducting uncover battles through well-trained agents, RAW has assumed a significant role in Indian foreign policy since its inception in 1968.

RAW's operations include the use of espionage, ethnic divisions, sabotage, terrorism and propaganda campaigns through the establishment of a massive network in target countries. The RAW leaders obtain critical information to satisfy Indian strategic interests and classify the data with the assistance of advanced technology and computer networks. Lately, RAW has been held accountable for being involved in Karachi and Balochistan unrest along with the assistance of foreign intelligence agencies.

RAW'S INVOLVEMENT IN KARACHI: CONFIRMED

More than 1 million weapons are imported to Karachi by the foreign intelligence agencies, contributing to violence and broader unrest in the area. The CIA, Mossad, and RAW are jointly working towards the ethnic division of Karachi by providing funds to military groups to conduct terrorist activities and destroy Pakistan's nuclear assets. Through the establishment of a collective network in Pakistan, the foreign spy agencies give the local population financial incentives, thus promoting their strategic interests in the area.

The poor, unemployed youth of Karachi is an easy target for the foreign secret agencies that recruit criminals, journalists, and security personnel and train young agents to execute targeted killings. Also, RAW supports Indian Muslims who join the Taliban regime in order to enhance religious schisms in Pakistan.

Pakistani Inter-Services Intelligence (ISI) has evidence of RAW's involvement in unrest in Karachi. After conducting terrorist activities, militants went to RAW's offices to collect their fee and get further directions. Friends and relatives of Maulana Fazlullah (a Pakistani Taliban) frequently meet with RAW leaders at the Indian Embassy in Afghanistan and collect money to support terrorist acts in Pakistan.

RAW AND CIA: CLOSE RELATIONS

Stephen P. Cohen of Brookings Institution claims that *"the CIA has assisted in the creation of RAW"* although their relations were established even before RAW was formed. In 1962, just after India's war with China ended, the CIA trained an undercover organization with Tibetan refugees in India to execute terrorist activities in China. On these grounds RAW was formed in 1968. And apparently, the relations of RAW and the CIA extend beyond the operational level.

Additionally, there has been a boost in India's undercover operations after 9/11 terrorist attack in New York. India became a strong supporter of the USA against the Muslim world and the USA opened the doors for India's intelligence services. The

CIAfacilitated the entry of RAW militants in Pakistan with the aim of destabilizing the region. Since 2008, targeted killings in Karachi and Balochistan have created the grounds for the emergence of a nationalistic movement, with the help of the Western allies, USA, UK and Germany, but also Afghanistan and Israel.

The oxymoron is that the CIA trained both agents of the ISI and RAW to use terrorism against an enemy. However, RAW's agents have received better and more sophisticated training than their Pakistani peers.

Bibliography

Das, Gupta, J.B. : *Islamic Fundamentalism and India* : Hope India Publications, Delhi, 2002.

Dhar, Maloy Kri. : *Mission to Pakistan: An Intelligence Agent in Pakistan,* Manas Publication, 2002.

Ganguly, Sumit : *"Avoiding War in Kashmir",* Foreign Affairs, New York, Winter 1990.

Multani, R K P : *Islamic Fundamentalism in South Asia* : Sumit Enterprises, Delhi, 2007.

Perkovich, George: *India's nuclear bomb: the impact on global proliferation.* University of California Press. 2002.

Raina, Asoka : *Inside RAW: The Story of India's Secret Service,* Vikas Publishing, New Delhi,1981.

Raj, Chengappa: *Weapons of Peace : the secret story of India's quest to be a nuclear power.* New Delhi: Harper Collins Publishers, India. 2000.

Sagar, Tariq Ismail : *RAW: Research and Analysis Wing,* Sagar Publication, 2007.

Saleem, Muhammad : *RAW: Global and Regional Ambitions,* Islamabad Policy Research Institute, Asia Printers, Islamabad, 2005

Shuja Nawaz: *Crossed Swords: Pakistan, Its Army, and the Wars Within,* Karachi: Oxford University Press, 2008.

Singh, Major General Sujan : *Phatoms of Chittagong; The Fifth Army in Bangladesh,* Allied Publishers, 1985, New Delhi.

Sumit Ganguly: *Conflict Unending: India-Pakistan Tensions since 1947,* New York: Columbia University Press, 2001.

Index

❑❑❑